Guide to Stability Design Criteria for Metal Structures

Guide to Stability Design Criteria for Metal Structures

Editor

Marek Lagunov

Guide to Stability Design Criteria for Metal Structures

Edited by **Marek Lagunov**

Printed in 2017

ISBN: 978-1-68117-146-3
Library of Congress Control Number: 2015954088

© 2016 by
SCITUS Academics LLC,
616, Corporate Way, Suite 2, 4766,
Valley Cottage, NY 10989

www.scitusacademics.com

This book contains information obtained from highly regarded resources. Copyright for individual articles remains with the authors as indicated. All chapters are distributed under the terms of the Creative Commons Attribution License, which permits unrestricted use, distribution, and reproduction in any medium, provided the original author and source are credited.

Notice

Reasonable efforts have been made to publish reliable data and views articulated in the chapters are those of the individual contributors, and not necessarily those of the editors or publishers. Editors or publishers are not responsible for the accuracy of the information in the published chapters or consequences of their use. The publisher believes no responsibility for any damage or grievance to the persons or property arising out of the use of any materials, instructions, methods or thoughts in the book. The editors and the publisher have attempted to trace the copyright holders of all material reproduced in this publication and apologize to copyright holders if permission has not been obtained. If any copyright holder has not been acknowledged, please write to us so we may rectify.

Preface

The Structural Stability Research Councilassist guidance to practicing engineers and writers of design specifications, codes, and standards in both posing simplified and refinedprocesses applicable to design and assessing their limitations.The main objectives of the Councilhave been tonurture research on the behavior of compressive components of metal structuresand of structural systems and to assist in the development of enhanced designprocedures.

This guide presents design of metal structurefor building and bridge design. It offers complete coverage of seismic connection design; cold metal framing connection; partially restrained connections; steel decks; inspection and quality control; and much more.

Guide to Stability Design Criteria for Metal Structures is a reference tool for consulting engineers, architects, building inspectors, and graduate students.

Table of Contents

Chapter 1: Methodology for the Design of Strength of Materials: Application to the Backfill of Narrow Trenches.............. 1

- ABSTRACT 1
- INTRODUCTION 1
- METHODOLOGY FOR THE DESIGN OF OPTIMISED CLSM 3
- NUMERICAL SIMULATION 5
  - Description of the Model 5
  - Materials 7
  - Variables Considered 9
  - Traffic Load 10
- RESULTS 11
  - Analysis of the Surrounding Soil 11
  - Analysis of the Pipeline 12
  - Analysis of the CLSM in the trench 13
  - Analysis of the Displacement in the Asphalt Pavement 14
- EXPERIMENTAL PROGRAM 15
  - Stage 1: Optimisation of the Aggregates 15
  - Materials and Mix Proportions 16
  - Stage 2: Optimisation of the Cement Content 19
- MATERIALS AND MIX PROPORTIONS 19
  - Tests and Results 20
  - Stage 3: Optimisation of Admixtures and Additions 20
  - Materials and Mix Proportions 21
  - Test and Results 21
- CONCLUSIONS 22
- ACKNOWLEDGEMENTS 22
- REFERENCES 23
- CITATION 24

Table of Contents

Chapter 2: Methodology for the Mix Design of Self-Compacting Concrete Materials Using Different Mineral Additions in Binary Blends of Powders... 25

ABSTRACT..25
INTRODUCTION 26
MATERIALS AND METHODS 30
   Materials..30
   Mix Proportions 32
   Test Program..41
TEST RESULTS AND DISCUSSION 43
   Fresh Properties...43
   Hardened Properties...59
   Proposed Methodology 61
CONCLUSIONS 63
REFERENCES 64
CITATION 68

Chapter 3: Reliability Design Methodology for Confined High Pressure Inflatable Structures....................................... 69

ABSTRACT..69
INTRODUCTION 70
   Limit States Design..73
   Determination of the Inflation Pressure...75
   Design for Material Strength 78
   Design for Axial Stability 80
MATERIAL CHARACTERIZATION 81
   Tensile Strength of the Fabric 82
   Friction Coefficient..85
PROTOTYPE DESIGN 88
   Inflation Pressure...88
   Fabric Strength..89
   Length of the Inflatable 89
SYSTEM RELIABILITY 90
   Inflation Pressure...91
   Axial Stability...92
   Material Strength 92
   Series System 92

Guide to Stability Design Criteria for Metal Structures

Influence of Importance Factor on System Reliability 93
CONCLUSIONS 98
ACKNOWLEDGEMENTS 98
REFERENCES 99
CITATION 100

Chapter 4:   Effect of Nitrite Ions on Steel Corrosion Induced by Chloride or Sulfate Ions... 101

ABSTRACT 101
INTRODUCTION 102
EXPERIMENTAL 104
RESULTS AND DISCUSSION 108
   E_{corr} Induced by Cl^- 112
   E_{corr} Induced by SO_4^{2-} 115
   Relationship between E_{corr} and NO_2^-/Cl^- 115
   Relationship between E_{corr} and NO_2^-/SO_4^{2-} 120
   Threshold Level of NO_2^-/Cl^- 122
   Threshold Level of NO_2^-/SO_4^{2-} 123
   Threshold Level of Cl^- 124
   Threshold Level of SO_4^{2-} 126
   Comparing the E_{corr} Induced by Cl^- and SO_4^{2-} 126
   Comparing the Threshold Level of Cl- and SO_4^{2-} 131
   Effect on Anodic/Cathodic Polarization Curves, Tafel Slope, and Stern-Geary Constant 134
CONCLUSIONS 147
REFERENCES 147
CITATION 150

Chapter 5:   Durability and Corrosion of Aluminium and Its Alloys: Overview, Property Space, Techniques and Developments 151

INTRODUCTION 151
   The General Performance of the Al-Alloy Classes 153
CORROSION OF ALUMINIUM AND ITS ALLOYS IN AQUEOUS ENVIRONMENT 157
   Environmental Corrosion of Aluminium 157
   Kinetic Stability of Aluminium Alloys 161
   The Property Space and Corrosion Property Profile of Aluminium Alloys 164

Table of Contents

CORROSION OF ALUMINIUM AND ITS ALLOYS IN AQUEOUS ENVIRONMENT....166
    The Role of Chemistry on Corrosion 166
    The Role of Microstructure on Corrosion 172
CORROSION PROTECTION 176
RECENT ADVANCES IN ASPECTS RELATED TO CORROSION OF ALUMINIUM
ALLOYS...182
    Formulation of Inhibitors:...183
    Evaluation in Aqueous Solution:...183
    Introduction into an Organic Matrix:..183
    Evaluation of the Performance of Organic Coatings:....................................184
    Optimisation: 184
HIGH THROUGHPUT ASSESSMENT..186
AC/DC/AC ACCELERATED TECHNIQUE FOR COATING EVALUATION 188
STAIRCASE IMPEDANCE...190
POTENTIOSTATIC TRANSIENTS FOR DETERMINATION OF METASTABLE AND
STABLE PITTING 191
SUMMARY AND CHALLENGES 194
REFERENCES 195
CITATION 212

Chapter 6: Finite Element Modeling of Steel Concrete Beam Considering Double Composite Action…………………….. 213

ABSTRACT...213
INTRODUCTION 214
RESEARCH SIGNIFICANCE 218
METHODOLOGY AND THE ANALYTICAL MODEL 218
MATERIAL PROPERTIES OF THE PROPOSED MODEL 226
VALIDATION OF THE ANALYTICAL MODEL 228
    Load–Deflection Relationship 229
    Interface Slip Values along the Beam Length 231
    Interface Slip Strain Values along the Beam Length 235
PARAMETRIC STUDIES...238
    The Influence of Removing the Lower Slab 239
    The Influence of Varying the Steel Beam Height 240
    The Influence of Varying Lower Slab Length..242
    The Influence of Varying Lower Slab Thickness 243
    The Influence of Varying the Head Studs Arrangement 245
    The Influence of Varying the Head Studs Diameter.....................................247

Guide to Stability Design Criteria for Metal Structures

CONCLUSIONS 248
REFERENCES 250
CITATION 252

Chapter 7:   Protection of Steel Corrosion in Concrete Members by the Combination of Galvanic Anode and Nitrite Penetration.. 253

INTRODUCTION 253
CHARACTERISTICS OF CORROSION PROGRESS OF STEEL BARS............................254
   Corrosion Condition of Steel Bars in the Bridge Slabs 254
   Experiments for Confirming the Formation of Macrocellsin Steel Bars 256
EFFECTIVENESS OF CORROSION INHIBITOR-CONTAINING MORTAR LAYERS 258
   Effects of Lithium Nitrite on the Corrosion of Steel Bars and the Movement of NO_2^- Ions from Mortars to Concrete. 258
   Quantitative Evaluation of Permeability of NO_2^- Ion.......................................260
CONFIRMATION OF THE EFFECTIVENESS OF ZINC WIRE AS A GALVANIC ANODE
....262
   Exposure Tests 262
PRACTICES IN THE COMPOSITE CORROSION PROTECTION METHOD 265
   Overview of the Method..265
   Details of Design 268
   A Case Study of the Application 269
SUMMARY AND CONCLUSIONS 276
   Conflict of Interests...276
REFERENCES 277
CITATION 277

Index... 279

Chapter 1

Methodology for the Design of Strength of Materials: Application to the Backfill of Narrow Trenches

A. Blanco, P. Pujadas, S.H.P. Cavalaro, A. Aguado

Department of Construction Engineering, UniversitatPolitècnica de Catalunya-BarcelonaTech, UPC, Jordi Girona 1-3, 08034 Barcelona, Spain

ABSTRACT

The design of controlled low-strength materials (CLSM) is generally based on experimental approaches without considering an efficient use of the component materials. The present study proposes a general methodology for the design of optimised CLSM that includes the definition of the mechanical requirements through numerical simulations with FEM and an experimental procedure to define the mix by optimising the aggregate skeleton, the content of cement and the use of admixtures and additions. Moreover, the methodology is applied to the backfill of narrow trenches.

INTRODUCTION

The use of narrow trenches for the installation of flexible pipelines of small diameter is a common technique for the construction of water, electricity, lighting and gas networks. One of the advantages of this technique is the limited interference with other services or traffic during construction. After the trench is excavated, a backfill material is used to fill the void left behind as well as to provide the support for the pipe and the surface elements. For that purpose, it is common to apply a controlled low-strength material (CLSM) as opposed to the use of traditional compacted granular fill.

As a backfill material, the CLSM requires a consistency close to that of self-compactability in order to reach tight or restricted-access areas [1] and a compressive strength that allows fast reestablishment of traffic without settling under traffic load, but also that may be easily excavated with conventional digging equipment [2]. These contradictory requirements represent a challenge when designing the mix proportions since both the deformability and the compressive strength of the material must be balanced and limited to a certain value.

In general, the definition of the mix proportions of CLSM are based on empirical approaches [3] and [4] that do not always consider the optimisation of the materials used. In order to develop an optimised CLSM, the design of the mix should be preceded by a technical base assessment of the requirements according to the application for what it is intended (e.g. backfills, structural fills, insulating or isolation fills, pavement bases, void filling, etc.). Afterwards, a series of tests should be performed to ensure that the resulting material fulfils the requirements.
Considering the abovementioned, the present study aims at defining a general methodology for the design of optimised CLSM. This approach proposes the assessment of the requirements through a numerical simulation of the application and an experimental procedure to optimise each of the components of the CLSM. Furthermore, this methodology is subsequently applied to a real case of a CLSM for the backfill of narrow trenches.

The main interest of the study consists in the development of a straightforward methodology that may be employed for any application. The novelty of the methodology is the use of a numerical analysis by means of FEM for defining the mechanical requirements of the CLSM since this type of studies are found in the literature for traditional compacted granular fill (e.g. for trenches [5] and [6]) but are scarce for

CLSM. Likewise, the methodology provides a series of tests for optimising the aggregate skeleton, the cement content and use of admixture that have proved to be convenient for the procedure.

METHODOLOGY FOR THE DESIGN OF OPTIMISED CLSM

The methodology proposed is summarized in the flow diagram presented in Fig. 1. The main requirements of a CLSM, according to what was previously exposed, are the workability in fresh-state and the compressive strength in the hardened-state. These two properties and, therefore, the requirements are dependant of the type of application for what the material is intended.

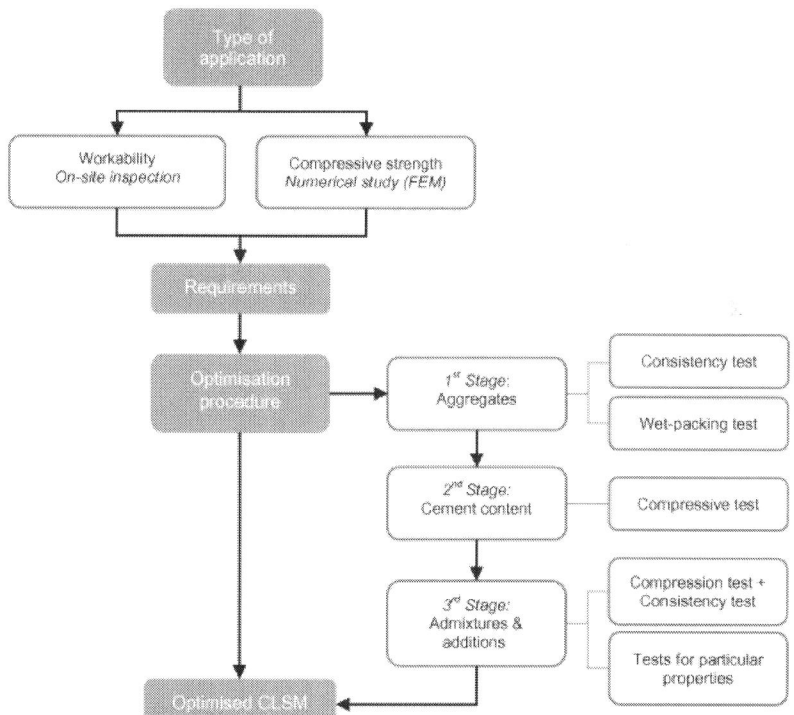

Figure 1: General methodology for the design of optimised CLSM.

In general, the workability of the CLSM should be close to self-compactability to avoid the need of vibration. Hence, the design of the mix should be conducted to attain that consistency. In order to set the value of compressive strength that materials should reach to fulfil the requirements, a FEM analysis is performed. This type of analysis allows reproducing the loading and boundary conditions of the CLSM and identify the compressive strength required to guarantee no settlements and easy excavation.

Afterwards, when the desired workability and compressive strength have been defined, the optimisation procedure of the mix is conducted in three stages: first the aggregate skeleton is optimised, subsequently the amount of cement and, finally, the use of admixtures and additions.
The optimisation of the aggregate skeleton consists in finding the combination of the aggregates that provides the maximum compactness and the desired workability. For that, two tests are defined at this stage: a wet-packing test [7] and [8] and the consistency test according to UNE-EN 1015-3:2000 [9].

The first one allows determining the solid concentration and the voids ratio, which are key parameters in the optimisation since a reduction of the voids could contribute to reduce the consumption of cement and problems regarding segregation. Furthermore, a more compact matrix would be less deformable, which is very convenient considering the applications of CLSM. The second one provides information regarding the consistency through the extent of the flow. Notice that a small content of cement (e.g. 40 kg/m3[7]) may be used to perform such tests given that the assessment of the strength is not in the scope of this stage.

The subsequent stage of the methodology is the optimisation of the amount of cement considering the aggregate skeleton obtained from the previous stage. For that purpose, the compression test according to UNE-EN 12390:2009 [10] is proposed to assess mixes with different cement contents and determine which is adequate to fulfil the requirements set previously.

Finally, the last stage of the methodology summarized in Fig. 1 is the optimisation of the use of admixtures and additions. Besides the traditional admixtures and additions (e.g. air-entraining admixtures, foaming agents or fly ash [2]), CLSM support the use of recycled or reused components from diverse origin such as recycled fine aggregates [11], waste materials, industrial by-products [12], [13], [14] and [15] and mine tailings [16].

In case any of the abovementioned materials are used, the consistency test and compression test previously proposed should be performed to assess the influence of such materials on the workability and compressive strength of the mix. Furthermore, if the additive or admixture has a particular property (e.g. expansive additions) this should be assessed to determine the effect on the fresh-state and hardened-state properties of the CLSM. In the case of an expansive addition, the resulting expansion should be measured.

The approach followed in this optimisation procedure is characterised by the use of water–solid ratios (w/s) instead of water– cement ratios (w/c). Notice that the terminology "solids" includes aggregates and cement. This provides a great flexibility in the sense that, for the value of w/s that leads to the workability desired, the cement content may be varied in order to increase or decrease the strength of the CLSM by replacing the equivalent amount of aggregates without affecting the workability. This is possible also due to the small content of cement used in this type of mixes in comparison with that usually found in conventional mortar. If the content of cement was an important part of the total volume, its influence in the consistency should be taken into account.

In subsequent sections, the methodology presented is applied to the case of a backfill material for a narrow trench for natural gas plastic pipelines.

NUMERICAL SIMULATION

Description of the Model
In order to determine the mechanical requirements of a CLSM for the backfill of a narrow trench, a numerical study is conducted with the finite

element software DIANA 9.1 [17]. This software is chosen to simulate the behaviour of the trench due to its extensive material library and analysis capabilities. A 3D model is considered and the four-node prismatic elements are used for the meshing, with a denser refinement in the trench and around the pipeline.

Fig. 2 depicts the geometry of the model and the location where the traffic load is applied (further detail on the definition and location of the load is given in subsequent sections). The geometry is defined according to the hypothesis that the surrounding soil may present layers with different properties or the constructive process adopted. Notice that the dimensions of the layers (both in depth and width) are defined based on the analysis of typical dimensions of narrow trenches found in practice and in the literature. Particularly, the width of the model (1.65 m) was defined so that it was representative of the actual situation and that its dimension did not affect the results of the simulation.

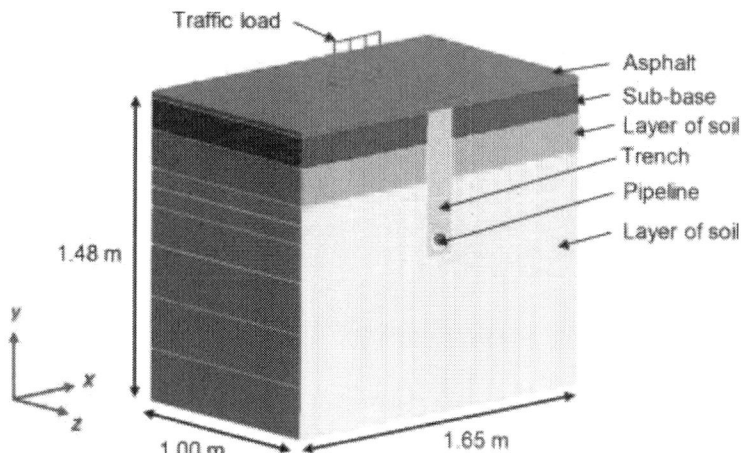

Figure 2: Geometry of the 3D model in DIANA 9.1.

Taking into account the abovementioned, the following layers and their corresponding depths are considered: an asphalt pavement of 0.03 m, a sub-base layer of 0.15 m, soil layer with a total depth of 1.30 m. Notice that the old and new asphalt pavement (after the trench is closed) overlap 0.10 m. The width of the narrow trench is 0.15 m and a depth of 0.70 m. On top of the trench there is the 0.03 m layer of asphalt pavement. The width of the surrounding soil at each side of the trench is

0.75 m and the total depth of the model, considering the pavement, the sub-base and the soil, is 1.48 m.

The diameter of the pipeline, which is simulated up to a length of 1.0 m, is 90 mm and lies on 0.03 m of CLSM. Notice that uniform bedding support under the pipeline was assumed in the simulation.

The boundary conditions set in the model correspond to a constraint support in axis z and an elastic support in axes x and y with a spring constant (k) of equal to 109.

Materials

The definition of the materials in the model is conducted based on the wide range of situations that may be found in practice when excavating a trench, such as several layers and types of soils. These situations in practice may lead to different requirements of the CLSMand, therefore, must be contemplated in the numerical study. For that reason, a parametric study is performed in which the variables are the properties of the soil and the CLSM (their properties are defined in the subsequent section). The properties of the asphalt pavement, the sub-base material and the pipeline remain constant given that their influence on the overall behaviour and the requirements of the CLSM are minor compared if compared to that of the soil since their characteristics are easier to control.

Fig. 3 shows an outline of the approach followed for the numerical simulations of the trench. Notice that the properties of the materials are the input for the model and are categorised in variables and data according to the previously mentioned.

8 | NUMERICAL SIMULATION

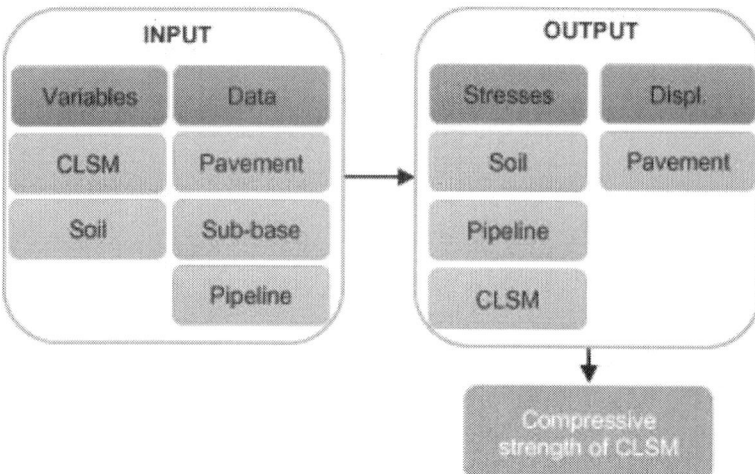

Figure 3: Approach followed in the numerical simulation of the trench.

Among the results obtained with the simulation, the analysis focus on the displacement in the asphalt pavement due to the traffic loads, the maximum stresses in the CLSM (or trench), in the soil and in the pipeline, which are represented in Fig. 3 as the output. Finally, based on the results a compressive strength for the CLSM is defined.

In general, the narrow trenches are excavated in pavements that have been used for several months or years, which indicates that the soil may be considered as consolidated. As a result, high values of stresses are not expected and, therefore, linear-elastic behaviour of the materials is considered.

For that reason, the properties of the materials are defined by the modulus of elasticity (E) and the Poisson coefficient (v) according to the literature [18]. For the material of the sub-base layer, a granular fill is considered and the pipeline is made of high density polyethylene (PE). Table 1 presents the values of E and v considered in each case.

Table 1: Properties of the materials considered as data.

Materials	Modulus of elasticity (MPa)	Poisson coefficient (-)
Asphalt pavement	3000	0.25
Sub-base material	350	0.35
Pipeline (PE100HD)	900	0.38

Two types of contacts between layers are defined in the model: a perfect contact and a weak contact. In the first case, for the horizontal contacts between the layers of asphalt pavement, sub-base material and soil and for the vertical contact between the old and new asphalt pavement a linear-elastic contact with a high stiffness is assumed (K = 1014 N/m). The second type corresponds to the vertical contact between the soil and the trench. Given the possible shrinkage of the CLSM, a weak contact is assumed by defining a low stiffness (K = 1 N/m).

Variables Considered

The variables considered in the numerical simulation are, as previously introduced, the properties of soil and the CLSM. Table 2 presents the properties the variables considered in the parametric study. The three types of soil were defined according to the literature [19]. The value of v remains constant since it is a common value representative of different types of soils. The properties of 5 types of CLSM were defined according to the values of E and v. Based on the literature [14], E may assume values ranging from 100 MPa to 300 MPa.

Table 2: Variables considered in the parametric study.

Variables		Modulus of elasticity (MPa)	Poisson coefficient (-)	Description
Soil	I	60	0.3	Hard clay, dense sand or loose gravel
	II	100	0.3	Hard clay, sandy clay or loose gravel
	III	200	0.3	Sandy clay or dense gravel
CLSM	I	100	0.2	–
	II	150	0.2	–
	III	200	0.2	–
	IV	250	0.2	–
	V	300	0.2	–

Besides simulating the behaviour of the trench with the three soils defined in Table 2, the possibility of having two different soils surrounding the trench is also studied. For that, an upper layer with the properties of soil I and a lower layer of soil II were modelled. In such case, the depths of the upper layer and the bottom layer are 0.30 m and 0.93 m, respectively. Therefore, the trench of 0.70 m of depth is surrounded by 0.15 m of the sub-base material, subsequently there is a layer of 0.30 m of soil I and the remaining depth of the trench is surrounded by soil II.

Traffic Load

The traffic load applied in the model is defined according to international traffic regulations, which specify the maximum load allowed per tyre. The regulations taken as a reference are the French and Spanish traffic regulations [20] and [21]. Considering the critical case of double tyres, the maximum load per axis specified is 130 kN in the former and 115 kN in the latter. To be on the safe side and to comply with both regulations, the highest value was used in all simulations.

To simplify the model and reduce the number of elements, the load is not applied to the surface of the tyre footprint. Instead, it is concentrated on the generatrix of the tyre (the width of the tyre is 300 mm), which also contributes to an assessment of the behaviour in a more critical situation that should be on the safe side.

RESULTS

Analysis of the Surrounding Soil

Fig. 4 depicts other results obtained from the analysis regarding the compressive stresses in the surrounding soil. The evolution of the compressive stresses with the modulus of elasticity of the CLSM is plotted for different types of soil. Notice that the case with two types of soil is referred to as (E = 60–100 MPa).

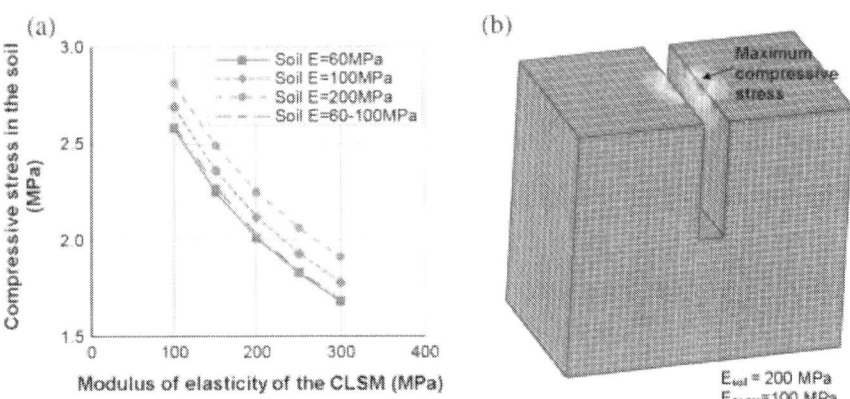

Figure 4: (a) Influence of the modulus of elasticity of the CLSM on the compressive stress and (b) stress distribution in the surrounding soil a soil of 200 MPa and CLSM of 100 MPa.

The curves in Fig. 4a reveal that the compressive stresses in the soil decrease with the modulus of elasticity of the CLSM due to the higher bearing capacity of the CLSM with bigger elastic modulus. For example, when the modulus of elasticity of CLSM increases from 100 MPa to 300 MPa the decrease of the stresses in the soil is 34.9% for a soil of 60 MPa and 32.0% for a soil of 200 MPa. These results reveal that the decrease observed is not significantly influenced by the type of soil.

Furthermore, the stresses are lower as the soil becomes more flexible. The increase of the stresses in the soil when it becomes stiffer is 40.9% for a modulus of the CLSM of 100 MPa. The highest stresses in the soil were obtained for stiff soil (E = 200 MPa) and a flexible CLSM (E = 100 MPa) and the compressive stress in the soil is 2.8 MPa. The results also indicate that the case with two types of soil (E = 60–100 MPa) are very similar to the ones obtained for the soil of 60 MPa. Hence, the response of the soil is governed by the top surface which is most flexible.

The distribution of stresses in Fig. 4b indicates that the maximum compressive stress is located at the boundary of the soil with the trench, particularly concentrated in the zone where the traffic load was defined (notice that the load was concentrated in the generatrix of the tyre, see Section 3.4 Traffic load).

Analysis of the Pipeline

For the case of the pipeline, presented in Fig. 5a, the trend is not as marked as for the soil. For flexible soils, the tendency of the compressive stresses is to decrease slightly, in particular an 8.6% for a soil of 60 MPa. However, for stiffer soils (e.g. E = 200 MPa), the compressive stresses in the pipeline increase slightly, thus leading to a difference between the most flexible CLSM (100 MPa) and the stiffest (300 MPa) of only 1.9%.

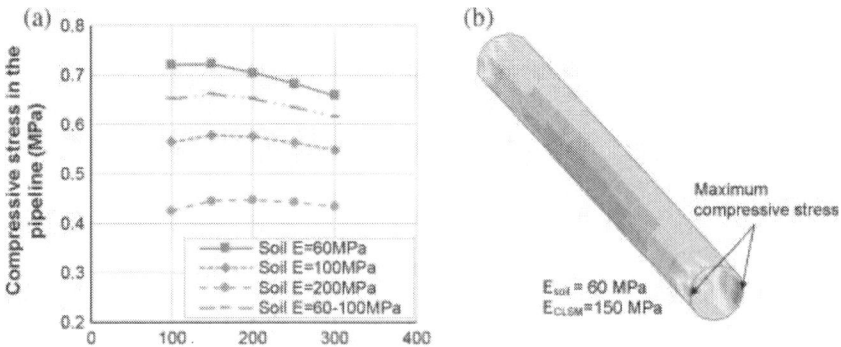

Figure 5: (a) Influence of the modulus of elasticity of the CLSM on the compressive stress and (b) stress distribution in the pipeline a soil of 60 MPa and CLSM of 150 MPa.

Guide to Stability Design Criteria for Metal Structures | 13

With regards to the modulus of elasticity of the soil, as the soil turns stiffer the stresses in the pipeline are smaller due to the higher bearing capacity of the soil. Notice that for a CLSM of 100 MPa, the compressive stress is reduced 40.9% when the modulus of elasticity of the soil increases from 60 MPa to 200 MPa. The biggest compressive stress (0.7 MPa) occurs for the more flexible soil and CLSM. Contrarily to what was observed in Fig. 4a, the case with two types of soils presents a noticeable difference with the other cases, exhibiting values of stress between those of the cases with only one layer of soil of 60 MPa and 100 MPa.

Fig. 5b shows the distribution of stresses along the pipeline. The maximum compressive stresses are observed at the ends of the pipeline and, particularly, in the haunches rather than at the top or the bottom.

Analysis of the CLSM in the trench

Fig. 6a presents the evolution of the maximum compressive stresses in the CLSM of the trench with the modulus of elasticity of the CLSM for different soil configurations. The curves reveal that the compressive stresses in the trench become higher as the modulus of elasticity of the CLSM increases. This outcome was expected since the increasing stiffness of the material with regards to the surrounding soil leads to the bearing of higher values of stress.

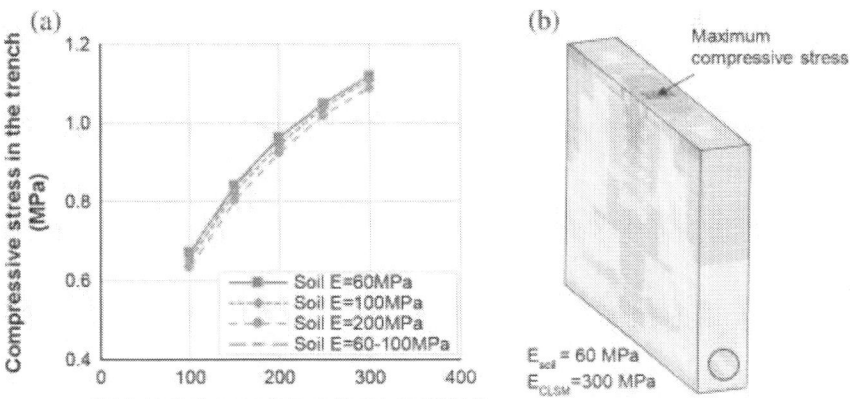

Figure 6: (a) Influence of the modulus of elasticity of the CLSM on the compressive stresses in the trench and (b) stress distribution in the trench for a soil of 60 MPa and CLSM of 300 MPa.

RESULTS

For a soil of 60 MPa, the increase in the modulus of elasticity of the CLSM from 100 MPa to 300 MPa represents an increase of the maximum compressive stress in the trench of 66.7% and for a soil of 200 MPa this percentage is 71.9%. Fig. 6b shows the stress distribution in the trench with the maximum compressive stress located at the top of the trench where the loading along the generatrix of the tyre was defined.

The worst-case scenario with regards to the requirements of compressive strength of the CLSM corresponds to the most flexible soil (E = 60 MPa) and the stiffest CLSM (E = 300 MPa). In this case the maximum compressive stress exhibited is 1.12 MPa.

The curves also reveal that the influence of the modulus of elasticity of the soil is very small since the differences observed between the softest and the stiffest soil are 6.0% for a CLSM of 100 MPa and 2.8% for a CLSM of 300 MPa. The reason for such small differences may be attributed to the fact that the maximum stress is located too close to the surface so that the influence of the surrounding soil is diminished.

Analysis of the Displacement in the Asphalt Pavement

The curves in Fig. 7 show the evolution of the displacement of the asphalt pavement with the modulus of elasticity of the CLSM. In this case, the opposite trend is observed since, as the modulus of elasticity of the CLSM increases, the displacement of the asphalt pavement becomes smaller due to the increasing stiffness of the CLSM.

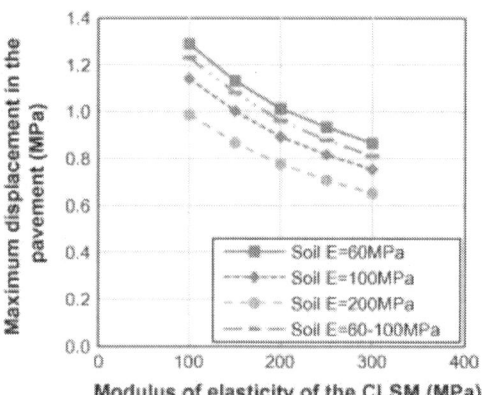

Figure 7: Influence of the modulus of elasticity of the CLSM on the compressive stress in the pipeline.

In addition to that, a noticeable influence of the properties of the soil is detected in the displacements estimated. For example, a soil with $E = 60$ MPa leads to a displacement 14.7% bigger than a soil with 100 MPa (for a CLSM of 300 MPa). The curves indicate that the configuration of two layers of soil (60–100 MPa) result in values of displacement that are in between of those obtained for one-layer soils of 60 MPa and 100 MPa. According to Fig. 6, the maximum displacement of the asphalt pavement occurs for the most flexible soil ($E = 60$ MPa) and CLSM ($E = 100$ MPa) with a value of 1.3 mm.

From the results it may be derived that the maximum compressive stresses due to the heavy traffic loading in service is 1.12 MPa. Consequently, a compressive strength of the CLSM equal to that value would be sufficient to bear the stresses and to guarantee a maximum displacement of 1.3 mm. Nevertheless, in order to avoid excessive micro-cracking (a non-linear regime) the maximum stresses in service are limited to a range of 45–55% of the compressive strength, leading to a strength ranging between 2.0 MPa and 2.5 MPa.

EXPERIMENTAL PROGRAM

The main objective of this section is to apply the experimental procedure of the methodology to obtain an optimised CLSM that attends the requirements set in the previous section. According to the optimisation procedure in the methodology this section is divided into three parts: the definition of the granular skeleton, the definition of the cement content and, finally, the definition of the additives and admixtures employed.

Stage 1: Optimisation of the Aggregates

At this stage, the influence of the granular skeleton on the workability and the packing density of the CLSM is assessed by considering different combinations of the aggregates and several water/solid ratios (w/s).

Materials and Mix Proportions

The aggregates used in the study are limestone sands of 0/2 and 0/4 with a density of 2510 kg/m3 and 2520 kg/m3, respectively, and a water absorption of 7.20% and 5.50%, respectively. They were combined in the percentages by volume of 80–20%, 50–50% and 20–80%, respectively. The values of w/s are 0.25, 0.30, 0.35, 0.40 and 0.45, assuming a situation of saturated surface-dry conditions (*SSD*). 1 Likewise, the cement content of reference at this first stage is of 40 kg/m3 (CEM II/A-M (V-L) 42.5 R), defined according to [5]. Table 3 presents the different mix compositions studied at this stage.

Table 3: Mix proportions in Stage 1.

w/s	Cement	Case 1: 80–20%			Case 2: 50-50%			Case 3: 20–80%		
		Sand 0/2	Sand 0/4	Water	Sand 0/2	Sand 0/4	Water	Sand 0/2	Sand 0/4	Water
0.25	40	1580	397	200	987	991	200	395	1586	200
0.30	40	1518	381	231	949	952	231	380	1524	231
0.35	40	1461	367	259	913	916	259	365	1467	259
0.40	40	1408	353	286	880	883	286	352	1413	286
0.45	40	1358	341	310	849	852	310	340	1364	310

All CLSM were manufactured according the same procedure. The solid components were first introduced in the 5 l mixer and then the water was added. These components were mixed during 1 min and the mixer was stopped for 30 s to ensure that material is not adhered to the surface of the recipient. Then, the mixer is started again and after 1 min the admixture is added. Finally, all the components are mixed for 2 min more.

Tests and Results

The tests conducted in this stage are the wet-packing test [7] and [8] and the consistency test according to UNE-EN 1015-3:2000 [9], as previously described in the methodology. The results of the former are presented in Fig. 8 in terms of the solid concentration and the voids ratio, which were calculated according to the formulations in [17].

Guide to Stability Design Criteria for Metal Structures | 17

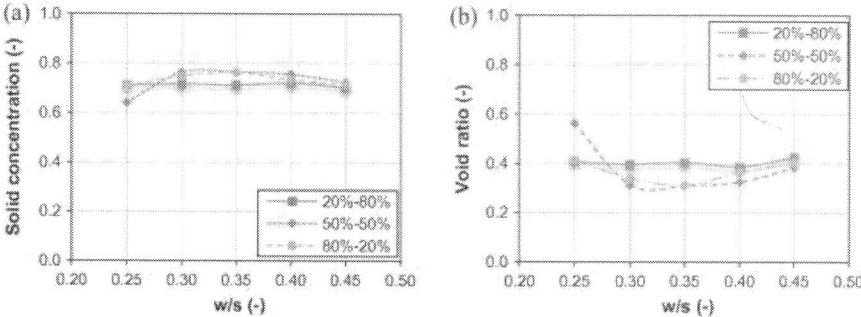

Figure 8: Influence of w/s on (a) solid concentration and (b) void ratio.

From the results in Fig. 8a and b it may be derived that the different combinations of aggregate skeleton exhibit, in general, similar values for the solid concentration and the void ratio. Such outcome indicates that, in this case, employing one or another combination of aggregates skeleton may not lead to significant differences in the results. The highest solid concentration and the lowest voids ratio is obtained for a w/s equal to 0.35, considering the granular skeletonsof 20–80% and of 500–50%. These values are 0.766 and 0.765 for the solid concentration, respectively, and 0.306 and 0.308 for the voids ratio, respectively.

Fig. 9 depicts the results of the consistency test in terms of the flow extent for different values of w/s considered. It reveals a clear influence of w/s in the consistency of the CLSM with an approximate S-shaped curve. The range of w/s values in which the increase of workability is significant, as observed due to the steep slope of the curve, is 0.30–0.40. For values higher than 0.40, the addition of water is less efficient and might lead to bleeding. Notice that the flow extent for the case of 800–20% with a w/s = 0.45 is not depicted in Fig. 9 since due to the fluidity of the mix it poured from the table.

18 | EXPERIMENTAL PROGRAM

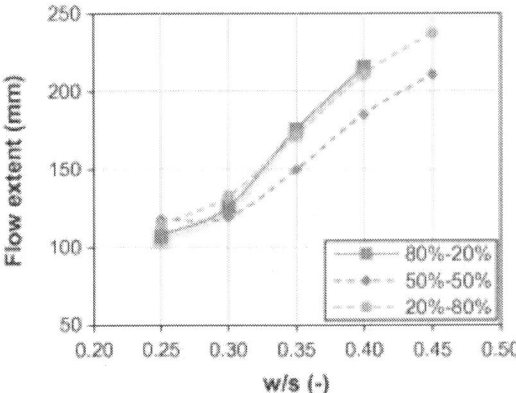

Figure 9: Influence of the w/s in the flow extent.

This phenomenon is due to the physical role of the water in the mortar and concrete mixes [22] and [23]. The aggregates in the mix require a certain amount of water to fill the pores (absorption water) and to wet the surface (wetting water). Notice that in the case of study the aggregates are already saturated. When the surface of the aggregate is wet, the remaining water has the function of fluidifying the mix (fluidification water) and, therefore, separating the particles. It should be highlighted that an excessive separation of the particles may lead to bleeding, a phenomenon that may be observed for high values of w/s.

From the values in Fig. 9, it may be stated that the threshold value to start fluidifying the mix is around W/S = 0.3. Fig. 10 shows the aspect of the CLSM with different W/S values after the consistency test and confirms the lack of fluidification water in the case of Fig. 10a.

Figure 10: Consistency test for 50–50% and (a) w/s = 0.25, (b) w/s = 0.35 and (c) w/s = 0.45.

Taking into account the results of the consistency test and the wet-packing test, it may be derived that a combination of the aggregates of

50% of sand 0/2 and 50% of sand 0/4 may be suitable for the application considered.

Stage 2: Optimisation of the Cement Content

The second stage of the experimental program focuses on the study of the minimum amount of cement that provides the minimum required compressive strength without compromising the excavability.

MATERIALS AND MIX PROPORTIONS

According to the ACI Committee 229 [1] the cement content in CLSM may vary between 30 and 120 kg/m3 depending on the requirements set. Considering this range, the cement contents assessed and presented in an internal report [24] ranged from 40 kg/m3 to 85 kg/m3. For this study, only the results of the content of cement that fulfil the strength requirements are shown. Notice that the w/s was set to 0.37 and the proportion of aggregates 50–50% was considered based on the results in Stage 1. The details of the mix proportions are included in Table 4.

Table 4: Mix proportions in Stage 2.

Materials		Quantities
w/s	(–)	0.37
Cement	(kg/m3)	75
Sand 0/2	(kg/m3)	885
Sand 0/4	(kg/m3)	882
Water	(kg/m3)	270

The mixing procedure of the components is the same as in the previous stage. In this case, 6 prismatic specimens of 40 × 40 × 160 mm were cast to determine the compressive strength following the specifications in UNE-EN 12390:2009 [19]. A total of 6 specimens were cast to be tested at 7 day, 21 days and 28 days.

Tests and Results

The test performed in the second stage is the compression test (UNE-EN 12390:2009 [19]). The results of the compression test at 7 days, 21 days and 28 days are presented in Table 5.

Table 5: Compressive strength of the CLSM with 75 kg/m3 for 7, 21 and 28 days.

	Average (MPa)	CV (%)
fc7	1.9	5.3
fc21	1.9	6.1
fc28	2.1	4.8

The results obtained indicate a compressive strength of 1.9 MPa both at 7 day and 21 days. The tests conducted at 28 days reveal a strength of the CLSM of 2.1 MPa. These values are in correspondence with the requirements set based on the results of the numerical model. Hence, a content of cement of 75 kg/m3guarantees the bearing of stresses resulting of the traffic load in the trench.

Stage 3: Optimisation of Admixtures and Additions

The final stage of the methodology studies the influence of the admixtures and additions on the properties of the CLSM. In this case, the experimental program only considers the influence of plasticizer (Pozzolith 475N) on the consistency and workability of the CLSM. However, the methodology may be applied to any additions by assessing their influence on the consistency, compressive strength and, if required, evaluating particular properties provided by the addition.

Materials and Mix Proportions

The mixes manufactured in the previous Section 4.1*Stage 1: Optimisation of the aggregates* were produced with 1.5% of plasticizer (in percentage of the cement weight). The other components of the mixes were added in the same proportions as in the previous case.

Test and Results

The assessment of the influence of the plasticizer was conducted by means of the consistency test according to the UNE-EN 1015- 3:2000 [17]. The curves showing the evolution of the diameter with w/s are depicted in Fig. 11.

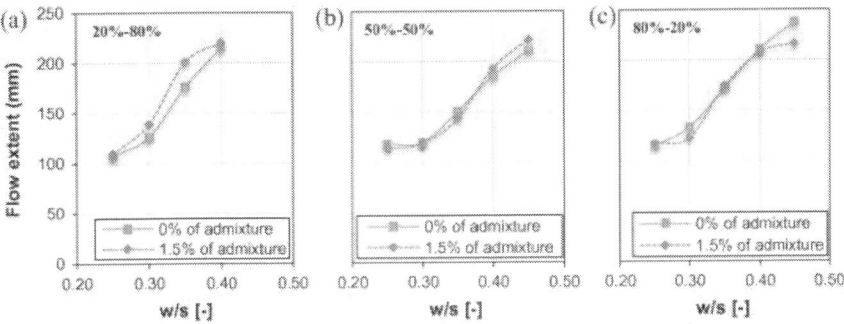

Figure 11: Influence of the admixture in the consistency of the CLSM for (a) 20–80%, (b) 50–50%, 80–20%.

The curves reveal that the influence of the plasticizer is almost unnoticeable on the consistency of the CLSM since the values of flow extent are very similar. Notice that the differences in average between the flow extent of the mixes without and with admixture are 7.2% for the case 20–80%, 0.8% for the case 50–50% and −3.5 for the case 80–20%.

In conventional concrete, the plasticizers produce a dispersion of the cement particles that otherwise would tend to form agglomerations and reduce the plasticity of the mix. In the case of CLSM, the small cement content and the high amount of water should reduce considerably the likelihood of particle interaction and flocculation. Therefore, the effect of the plasticizers should be less noticeable in CLSM.

CONCLUSIONS

The study conducted represents a contribution to the design of CLSM with optimised mix proportions. Furthermore, the methodology proposed allows a technical base assessment of the requirements associated to the application and provides the tests required to optimise the mix while fulfiling the requirements set. The following conclusions may be derived from the present study:

- The results of the numerical model confirm that a compressive strength ranging between 2.0 MPa and 2.5 MPa would ensure stability and excavability.
- The critical zones of loading in the surrounding soil and in the CLSM of the trench were identified.
- The load in the pipeline is much smaller and would only be a problem if it was transmitted through point-like aggregates.
- The consistency tests performed in Stage 1 of the experimental procedure for the optimisation of the mix revealed a range of values of w/s (0.35–0.40) leading to a suitable workability.
- The compression tests performed in Stage 2 reveal that a content of cement of 75 kg/m3 is enough to fulfil the strength requirements obtained from the model.
- The use of plasticizers to improve the workability was found to be inefficient for this type of material with low contents of cement, based on the results obtained in the test conducted in Stage 3. The methodology was successfully applied to the backfill of narrow trenches but could be also employed to design other types of CLSM that may include new materials (e.g. waste materials or industrial by-products) and to different applications.

ACKNOWLEDGEMENTS

The authors of this paper want to acknowledge the financial support provided by Gas Natural Fenosa (GNF) and, in particular, the comments and conversations with Jordi Rosselló and Carlos Aranda.

REFERENCES

1. Siddique R. Utilization of waste materials and by-products in producing controlled low-strength materials. ResourConservRecy 2009;54:1–8.
2. ACI Committee 229. Controlled low-strength materials (ACI 229R- 99). Farmington Hills, MI: American Concrete Institute; 1999.
3. Federal Highway Administration. User guidelines for waste and by-product materials in pavement construction. U.S. DOT. Report No. FHWA-RD-97-148; 1997.
4. Alizadeh A, Helwany S, Ghorbanpoor A, Sobolev K. Design and application of controlled low strength materials as a structural fill. Constr Build Mater 2014;53:425–31.
5. Blakaya M, Moore ID, Sagˇlamer A. Study of non-uniform bedding due to voids under jointed PVC water distribution pipes. GeotextGeomembranes 2012;35:99–108.
6. Balkaya M, Moore ID, Sagˇlamer A. Study of non-uniform bedding support under continuous PVC water distribution pipes. TunnUndergrSpTechnolm 2013;35:99–108.
7. Fung WWS, Kwan AKH, Wong HHC. Wet packing of crushed rock fine aggregate. Mater Struct 2009;42:631–43.
8. Wong HHC, Kwan AKH. Packing density of cementitious materials: Part 1 – Measurement using a wet packing method. Mater Struct 2008;41:689–701.
9. AENOR.UNE-EN 1015-3:2000 Métodos de ensayo para morteros de albañilería. Parte 3: Determinación de la consistenciadelmortero fresco (por la mesa de sacudidas), Asociación Española de Normalización y Certificación. Madrid; 2005.
10. AENOR. UNE-EN 12390:2009 Ensayos de hormigónendurecido. Parte 3: Determinación de la resistencia a compresión de probetas, Asociación Española de Normalización y Certificación. Madrid; 2009.
11. Etxebarria M, Ainchil J, Pérez ME, González A. Use of recycled fine aggregates for Control Low Strength Materials (CLSMs) production. Constr Build Mater 2013;44:142–8.
12. Taha RS, Alnuaimi AS, Al-Jabri KS, Al-Harthy AS. Evaluation of controlled low strength materials containing industrial by-products. Build Environ 2007;42:3366–72.
13. Nataraja MC, Nalanda Y. Performance of industrial by-products in controlled low-strength materials (CLSM). Waste Manage 2008;28:1168–81.
14. Wang H-Y, Chen B-T, Wu Y-W. A study of the fresh properties of controlled low-strength rubber lightweight aggregate concrete (CLSRLC). Constr Build Mater 2013;41:526–31.

15. Lee NK, Kim HK, Park IS, Lee HK. Alkali-activated, cementless, controlled lowstrength materials (CLSM) utilizing industrial by-products. Constr Build Mater 2013;49:738–46.
16. Bouzalakos S, Dudeney AWL, Chan BKC. Formulating and optimizing the compressive strength of controlled low-strength materials containing mine tailings by mixture design and response surface methods. Miner Eng 2013;53:48–56.
17. TNO Diana BV. Diana user's manual. 2008. <http://www.tnodiana.com/>.
18. Kraemer C, Pardillo JM, Rocci S, Romana MG, Blanco VS, del Val MA. Ingeniería de carreteras, vol. II. 1st ed. Madrid: McGraw-Hill Interamericana de España, S.A.U; 2003.
19. Jiménez Salas JA, Justo Alpañés JL. Geotencia y cimientos I. 2nd ed. Madrid: Rueda; 1975.
20. España. Real Decreto 2822/1998, de 23 de diciembre, por el que se aprueba el Reglamento General delVehículos. BoletínOficialdel Estado, núm. 22, 26 de enero de 1999. p. 3484–47.
21. France. Loi n 55-435 du 18 avril 1955 portantstatut des routes.
22. Fennis SAAM. Design o f ecological concrete by particle packing optimization [Doctoral Thesis]. Delft: Delft University of Technology; 2011.
23. Klein N. El Rolfísicodelaguaenmezclas de cemento Portland [Doctoral Thesis]. Barcelona: UniversitatPolitècnica de Catalunya; 2012.
24. Aguado A, Blanco A, Cavalaro S, Pujadas P. Desarrollo de mortero de relleno para zanjas de dimensionesreducidas. Internal Report, UniversitatPolitècnica de Catalunya. Barcelona; 2013.

CITATION

A. Blanco , P. Pujadas, S.H.P. Cavalaro, A. Aguado, Methodology For The Design Of Strength Of Materials: Application To The Backfill Of Narrow Trenches, http://dx.doi.org/10.1016/j.conbuildmat.2014.09.008

Chapter 2

Methodology for the Mix Design of Self-Compacting Concrete Materials Using Different Mineral Additions in Binary Blends of Powders

Miguel C.S. Nepomuceno[1], L.A. Pereira-de-Oliveira[1], S.M.R. Lopes[2]

[1]University of Beira Interior, Centre of Materials and Building Technologies, Portugal
[2]University of Coimbra, CEMUC, Portugal

ABSTRACT

Objective: Interaction between the coarse aggregates and the mortar phase of self-compacting concrete (SCC) was evaluated in a two phase program.

Materials and methods: In the first phase, 74 mortars suitable for SCC were produced, combining different volumetric ratios between powders and fine aggregates and different binary blends of powders. In the second phase, 60 concretes were produced with different volumetric ratios between the mortar phase and the coarse aggregates, and their fresh and hardened properties were evaluated.

Results: Based on this study, correlations between mix design parameters, fresh and hardened properties were obtained and a methodology was proposed for the mix design of SCC.

INTRODUCTION

Since the first developments of self-compacting concrete (SCC), several methods have been proposed for its mix design, especially in the 90s, coinciding to the period of a quick increasing use of SCC. In general, all methods reflect, with greater or lesser extent, some concern with the optimisation of the granular skeleton and the reduction of the paste volume. In large scale applications, the economy and robustness of the mixtures are always a present concern. It is a common ground that there is no single universal solution, but rather a very wide range of possible solutions that, taking into account locally available materials, are able to give a satisfactory outcome in each particular situation.

The first well known mix design method was proposed in 1993 by Okamura, Ozawa and Maekawa [1],[2] and [3], which was later improved in 1998 by the contribution of Ouchi et al. [4]. This method, developed in the University of Tokyo, was then known as the general method. The general method assumes as the starting point the design of the mortar phase, which must meet certain flow requirements, necessary to achieve a SCC. In the mortar phase, the ratio between the volume of fine aggregate and the volume of the mortar excluding air (V_s/V_m) is 0.40, and the water/powder ratio by volume (V_w/V_p) and superplasticizer/powder ratio by mass (Sp/P) are adjusted to obtain the required flow properties. The volume of coarse aggregates (V_g), with a maximum size of 20 mm, is calculated on the basis of 50% of dry compacted bulk volume of coarse aggregate excluding air content (V_{ap}). The slump-flow, v-funnel and U-box tests are used to evaluate the self-compactability of concrete. The general method considers some of the mix design parameters as almost constant, which allows little flexibility in optimising the granular skeleton, usually leading to higher portions of paste when compared to an optimised granular structure [5]. Afterwards, the general tendency was to focus on optimising mixture proportions, aiming to reduce paste volume, mainly by increasing the volume of aggregates. Some contributions to improve the general method were proposed by Pelova et al. [6], Domone et al. [7] and Edamatsu et al. [8]. The general idea was that it was possible to increase the volumetric ratio between the fine aggregate and mortar (Vs/Vm) and increase the volume of coarse aggregate (Vg) by reducing its maximum dimension, but the equilibrium between both variables is a focal point to achieve the self-compactability. A different approach from that used in the general method was developed in CBI by Petersson, Billberg and Bui [9] and [10] in 1996. They based their work on previous studies conducted by Bui and Tangtermsirikul and Bui [11]. In the CBI method, the concrete was assumed to be consisted basically of two phases, namely, the liquid phase (paste) and the solid phase (coarse and fine aggregates). The reduction of the paste volume is achieved by optimising the solid phase, combining the maximum inter-particle distance criterion and the blocking criterion. The first allows estimating the optimal coarse/fine aggregate ratio, while the second allows estimating the maximum volume of aggregates. The model includes the external conditions, such as the diameter and the spacing between steel bars. Later developments by Bui and Montgomery [12] considered the introduction of an additional criterion, called the criterion of liquid phase. The liquid phase criterion, in conjunction with the blocking criterion, leads to the evaluation of the minimum volume of paste required to produce a satisfactory SCC and to ensure proper passing ability in L-box test.

The method developed by Sedran and Larrard [5] in the LCPC has also represented a contribution to optimisation of the solid skeleton of the SCC. Its main feature is based on the use of a mathematical model developed by LCPC and called Compressive Packing Model (CPM). This model considers any range of sizes of materials and differs from the previous in that the cement, additions and aggregates are all included in the CPM. The method is supported on mathematical models that estimate the fresh properties as a function of the characteristics of materials and external conditions, such as the border effect provoked by formwork or the effect of spacing and diameter of the steel bars.

Many other approaches were developed, as for example the one proposed by Su et al. [13] and Su and Miao [14] for medium strength SCC, that starts by determining the volume of aggregates using the packing factor (PF) and only later the properties of the paste. Sonebi [15] has also investigated the effects of the cement content, additions and SP on the fresh and hardened properties of SCC and proposed a statistical model to simplify the test protocol required to optimise a given mix. Improving performance and robustness of SCC in large scale production emerged naturally as a key factor concern. Kwan and Ng [16] argue that decreasing the W/C ratio and increasing the fine/total aggregate ratio are both effective means to improve the performance and robustness of SCC. Kwan and Ng [17] have also confirmed that the addition of pulverized fuel ash and/or condensed silica fume can significantly improve the performance and, more importantly, the robustness of SCC.

The EFNARC specification and guidelines for SCC [18], published in 2002, reflects the so far practical experience and the latest research findings, and provided to his members in Europe a framework for design and use of high quality SCC. Indicative typical ranges of proportions and quantities are given for initial composition, assuming that further modifications could be necessary to meet strength and other performance requirements. The initial parameters are: a water/ powder ratio by volume (V_w/V_p) of 0.80–1.10; a total powder content (V_p) of 0.16–0.24 m3/m3 (400–600 kg/m3); a coarse aggregate content (V_g) of 0.280.35 m3/m3; a W/C ratio selected based on EN 206- 1:2000 [19] and a water content not exceeding 200 l/m3. In 2005, the European Guidelines for SCC was published [20]. This document did not proposed any standard method for SCC mix design, because, as it is mentioned, many academic institutions, admixture, ready mixed, precast and contracting companies have developed their own mix proportioning methods. The option was to establish the mix design principles based on general recommendations. In America, ACI [21] recommends different contents of ultrafine material from 355 kg/m3 to more than 458 kg/m3 depending on the required slump-flow, a volume of paste (V_{pw}) from 0.34 to 0.40 m3/m3, a volume of mortar (V_m) from 0.68 to 0.72 m3/m3, a W/C ratio by mass from 0.32 to 0.45 and a content of cement from 386 to 475 kg/m3.

Looking at the application of the SCC from 1993 to 2003, Domone [22] concluded that successfully performed SCC were obtained for a great variety of constituents and mix proportions, but considerable scope for optimisation of mixes for greater efficiency and economy was still possible. As the most critical parameters for successful SCC mix design, Domone [22] has identified: the coarse aggregate volume, the paste content of concrete and the fine aggregate percentage of the mortar. The powder content and water/powder ratio have shown greater flexibility. Median values of the key mix proportions were a coarse aggregate content (Vg) of 0.312 m3/m3, a paste content (Vpw) of 0.348 m3/m3, a powder content of 500 kg/m3, a water/powder ratio by weight (W/P) of 0.34, and a fine aggregate/mortar ratio (Vs/Vm) of 0.475. The selection of the component materials seems to depend on local availability, but some predominant features can be identified: the crushed rock with a maximum size between 16 and 20 mm was predominant; nearly all cases used either a binary or ternary blend of Portland cement with additions of all the types used in conventional concrete, but the limestone was the most common addition; all mixes included a SP, but to make mixtures more robust, almost half cases use a VMA in addition to SP. The filling ability of the concrete was evaluated mainly by the slump-flow and flow rate values and in some cases the L-box and the U-box tests were used to evaluate the passing ability. Slump-flow varied mainly in the range of 600–750 mm, v-funnel times varied from 3 to 15 s, L-box passing ratio values were all in excess of 0.8 and U-box filling height values were in excess of 300 mm, with the reinforcement spacing varied in some cases to suit the application. The 28 day compressive strength varied from 20 to nearly 100 MPa.

Despite the time elapse of about two decades after the first proposal for the mix design of SCC, the attempted for the optimisation of SCC mortar phase still carried on. Li and Kwan [23] considered that the rheology of a fresh concrete was largely determined by the rheology of its mortar portion and hence proper design of the mortar portion should be the first step in the mix design of concrete. Li and Kwan [23] have demonstrated that the factors affecting the rheology of cement paste include the water content, packing density and solid surface area, and that the combined effects of these factors may be evaluated in terms of the water film thickness (WFT). As a result, Li and Kwan [23] extended this concept to cement–sand mortar and purposed a mix design method based on the WFT. They found that both the WFT and cement/aggregate ratio have major effects on the rheology of mortar, but the WFT is still the single most important factor. More recently, Kwan and Li [24], found that the WFT play an important role in the adhesiveness and strength of mortar, and is therefore a key parameter to be considered in the design of mortar and concrete mixes.

Nepomuceno et al. [25] and [26] have also proposed a methodology for the mix design of the mortar phase of SCC that allows to reconcile fresh properties and compressive strength when binary blends of powders are used. Later, a new methodology for the mix design of SCC was proposed by the same research team[27]. This new methodology has been used in Portugal in recent years with success, and there are already few researches attesting its validity to different types of materials [28], [29] and [30], including recycled coarse aggregates [31] and [32]. Furthermore, with some adjustments its use can also be extended to concrete with lightweight aggregates [33] and [34]. In this sense, the authors have proposed to describe in this article the study that supported the proposal of this new methodology for the mix design of SCC.

The new methodology here described is based on simple procedures and assumes the SCC as a two phase material, the mortar phase and the coarse aggregates. The study was focused on SCC in which adequate viscosity is achieved by controlling the amount of powders. For concrete mix design, only binary blends of powders were used, including two cements and three additions: limestone powder, fly ash and granite filler. The granite filler originating from crushing of ornamental rocks, without any additional processing, was used on a trial basis [26]. Additions from industrial waste materials are being tested for use as filler in SCC, such as granite filler or marble dust [35] and [36], as it can provide economic benefits and prevent environmental pollution [36]. The parameters used for the SCC mix

MATERIALS AND METHODS

Materials

For this study, two cements were selected in accordance with NP EN 197-1 [37]: a Portland cement of type CEM I 42.5R (C1) with a density of 3140 kg/m3 and a limestone Portland cement of type CEM II/B-L32.5N (C2) with a density of 3040 kg/m3. The mineral additions included a limestone powder (FC) with a density of 2720 kg/m3, fly ash (CV) with a density of 2380 kg/m3 and granite filler (FG) with a density of 2650 kg/m3. More details concerning the fineness of the powder materials and the particle distribution analysis of mineral additions can be found elsewhere [26]. A polycarboxylate-based superplasticizer (SP) was used, having a density of 1050 kg/m3.

Two fine aggregates from natural origin were selected and identified as Sand 01 and Sand 05. Sand 01 (S1) was classified as fine sand, with a density of 2590 kg/m3, bulk density of 1490 kg/m3 and a fineness modulus of 1.49. The Sand 05 (S2) was classified as river sand, with a density of 2610 kg/m3, bulk density of 1570 kg/m3 and a fineness modulus of 2.71. The proportion between the two sands was determined experimentally in order to obtain the highest degree of packing of the granular skeleton in a bulk volume, resulting in 40% for Sand 01 and 60% for Sand 05, in percentage of the absolute volume of total fine aggregate (Vs). The obtained mixture was used as a reference curve for the mortar mix design, presenting a bulk density of 1598 kg/m3 and a fineness modulus of 2.22. The grading curves of fine aggregates and the resulting reference curve were presented elsewhere [26].

Two coarse aggregates from crushed granite were selected and identified as CA 3/6 and CA 6/15. The CA 3/6 (G1) had a density of 2710 kg/m3, bulk density of 1520 kg/m3 and a fineness modulus of 5.08. The CA 6/15 (G2) had a density of 2700 kg/m3, bulk density of 1540 kg/m3 and a fineness modulus of 6.47. The proportion between the two coarse aggregates was determined experimentally in order to obtain the highest degree of packing of the granular skeleton in a bulk volume, resulting in 50% for CA 3/6 and 50% for CA 6/15, in percentage of the absolute volume of total coarse aggregate (Vg). The obtained mixture was used as a reference curve for the concrete mix design, presenting a bulk density of 1642 kg/m3 and a fineness modulus of 5.78. The grading curves of coarse aggregates and the resulting reference curve are presented in Figure 1.

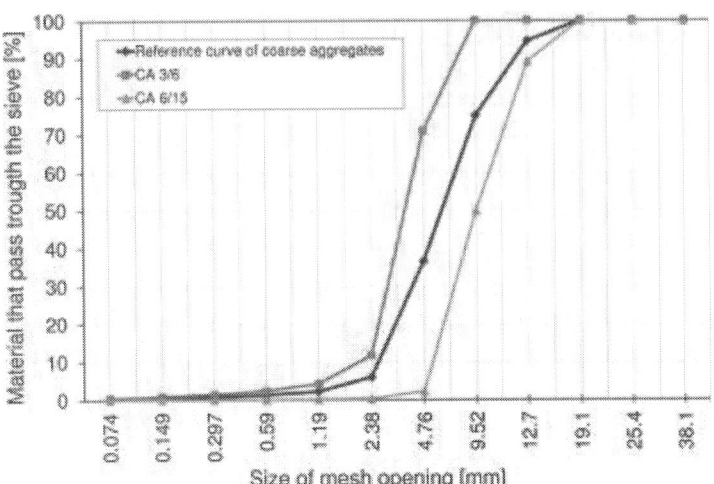

Figure 1: Grading curves of coarse aggregates and reference curve.

Mix Proportions

In the study of the mortar phase, the objective was to identify the relevance of the proposed mix design parameters and to define a methodology to simultaneously achieve the adequate fresh and hardened properties to successfully produce SCC. For this propose 15 binary blends of powder materials were produced to study the effect of powder mixture. The percentage of replacement of cement by the additions varied from 0% to 60%. To evaluate the effect of varying proportions between the volume of powder materials (Vp) and the volume of fine aggregates (Vs) in mortars, five different ratios of Vp/Vs were produced (0.60, 0.65, 0.70, 0.75 and 0.80).

All of the 74 produced mortars presented similar flow properties, represented by a slump-flow between 251 and 263 mm and a v-funnel time between 7.69 and 8.77 s. By simplification, in this paper, the adequate rheological properties of mortars and concretes should be interpreted as the adequate filling ability, measured indirectly by the slump-flow and flow rate. It is known that, under certain conditions of flow, without segregation and blocking, those parameters are related to the Bingham constants of yield stress and plastic viscosity, respectively. The adequate rheological properties were defined as an interval, based on bibliographic review and exploratory studies [26]. To achieve these rheological properties, an experimental procedure was defined to determine the adequate values of Vw/Vp and Sp/p% in mortars. This would allow evaluating the relationship between those ratios and the required properties in fresh and hardened state. The volume of voids when calculating mortars and the contribution towards volume of powder materials originating from fine aggregates were both overlooked. This study concerns to the first phase of the project and includes 74 different mortars, with similar rheological properties, different blends of powders and different proportions between powders and fine aggregates. The mix proportions were presented in Ref. [26].

In the study of the concrete phase, two objectives were established. The first was to confirm the adequacy of the proposed mortar flow properties to produce successfully SCC mixtures. The second, and more important objective, was to evaluate the influence of the coarse aggregate volume in workability properties of SCC mixtures and to define a methodology to successfully produce SCC. The parameters for the mix design of concrete have included the volume of voids (Vv), which was assumed as constant (0.03 m3) for concrete without air entraining admixtures, the ratio Vm/Vg (ratio between the absolute volume of mortar and the absolute volume of coarse aggregate), the volumetric ratios of each coarse aggregate (g1, g2, ... gn) to the total absolute volume of coarse aggregates (Vg) and all the parameters previously defined for the mortar phase.

A total number of 60 concretes were produced, using different mortars produced on the first phase as a matrix. The selection included different binary blends of powder materials, different Vp/Vs ratios (0.60, 0.70 and 0.80) and a wide range of mortar compressive strengths, ranging from 25 to 95 MPa. Each of the selected mortar matrixes was used to produce 4 different concretes, corresponding to 4 different coarse aggregate volumes, expressed by 4 different Vm/Vg ratios (2.0, 2.2, 2.4 and 2.6). The Vm/Vg ratios were selected is such a way that the volume of coarse aggregate (Vg) varied between the limits usually acceptable for SCC. As a result, different combinations between the Vp/Vs and the Vm/Vg ratios were obtained, which enabled evaluating the interaction between the coarse aggregates and the mortar matrix particles, with different materials and proportions between powders and fine aggregates. The interaction between those two ratios (Vp/Vs and Vm/Vg) was expressed by a parameter which was named MN (Mixture Number), and represents

MATERIALS AND METHODS

the product of Vp/Vs by Vm/Vg. The MN parameter was found to have a good and linear correlation to the total volume of aggregates and, indirectly, to the total volume of paste in concrete. The mix design parameters of all the concrete produced are presented in Table 1 and the mix proportions per cubic metre are shown in Table 2.

Table 1: Mix design parameters of SCC

Id	Mortar phase mix design parameters				Concrete phase		Relevant parameters for comparison to other methods					
	Blend of powders	Vp/Vs	Vw/Vp	Sp/p%	Vm/Vg	Vv (m³)	W/C	W/CM	Vs/Vm	Vp (m³)	Vg (m³)	Vap (%)
B.2.1	80C2+20FC	0.80	0.76	2.10	2.40	0.03	0.313	0.313	0.407	0.223	0.285	48.4
B.2.2	80C2+20FC	0.80	0.76	2.10	2.20	0.03	0.313	0.313	0.407	0.217	0.303	51.5
B.2.3	80C2+20FC	0.80	0.76	2.10	2.00	0.03	0.313	0.313	0.407	0.211	0.323	54.9
B.2.4	80C2+20FC	0.70	0.81	2.10	2.40	0.03	0.333	0.333	0.433	0.208	0.285	48.4
B.2.5	80C2+20FC	0.70	0.81	2.10	2.20	0.03	0.333	0.333	0.433	0.202	0.303	51.5
B.2.6A	80C2+20FC	0.70	0.81	2.10	2.00	0.03	0.333	0.333	0.433	0.196	0.323	54.9
B.2.6B	80C2+20FC	0.70	0.81	2.10	2.00	0.03	0.333	0.333	0.433	0.196	0.323	54.9
B.2.7	80C2+20FC	0.60	0.90	2.05	2.40	0.03	0.370	0.370	0.460	0.189	0.285	48.4
B.2.8	80C2+20FC	0.60	0.90	2.05	2.20	0.03	0.370	0.370	0.460	0.184	0.303	51.5
B.2.9	80C2+20FC	0.60	0.90	2.05	2.60	0.03	0.370	0.370	0.460	0.193	0.269	45.7
B.2.10	80C2+20FC	0.70	0.81	2.10	2.60	0.03	0.333	0.333	0.433	0.212	0.269	45.7

Guide to Stability Design Criteria for Metal Structures

B3.1	80C2 + 20FG	0.80	0.82	2.15	2.40	0.03	0.337	0.337	0.399	0.219	0.285	48.4
B3.2	80C2 + 20FG	0.80	0.82	2.15	2.20	0.03	0.337	0.337	0.399	0.213	0.303	51.5
B3.3	80C2 + 20FG	0.80	0.82	2.15	2.00	0.03	0.337	0.337	0.399	0.207	0.323	54.9
B3.4	80C2 + 20FG	0.70	0.87	2.15	2.40	0.03	0.358	0.358	0.425	0.204	0.285	48.4
B3.5	80C2 + 20FG	0.70	0.87	2.15	2.20	0.03	0.358	0.358	0.425	0.199	0.303	51.5
B3.6	80C2 + 20FG	0.70	0.87	2.15	2.00	0.03	0.358	0.358	0.425	0.193	0.323	54.9
B3.7	80C2 + 20FG	0.70	0.95	2.10	2.40	0.03	0.391	0.391	0.453	0.186	0.285	48.4
B3.9	80C2 + 20FG	0.60	0.95	2.10	2.60	0.03	0.391	0.391	0.453	0.191	0.269	45.7
B3.10	80C2 + 20FG	0.70	0.87	2.15	2.60	0.03	0.358	0.358	0.425	0.209	0.269	45.7
B4.1	80C2 + 20CV	0.80	0.76	2.10	2.40	0.03	0.312	0.261	0.407	0.223	0.285	48.4
B4.3	80C2 + 20CV	0.80	0.76	2.10	2.00	0.03	0.312	0.261	0.407	0.211	0.323	54.9
B4.4	80C2 + 20CV	0.70	0.81	2.10	2.40	0.03	0.333	0.279	0.433	0.208	0.285	48.4
B4.5	80C2 + 20CV	0.70	0.81	2.10	2.20	0.03	0.333	0.279	0.433	0.202	0.303	51.5
B4.7	80C2 + 20CV	0.60	0.89	2.05	2.40	0.03	0.366	0.306	0.461	0.189	0.285	48.4

MATERIALS AND METHODS

Sample	Mix											
B.4.9	80C2+20CV	0.60	0.89	2.05	2.60	0.03	0.366	0.306	0.461	0.194	0.269	45.7
B.4.10	80C2+20CV	0.70	0.81	2.10	2.60	0.03	0.333	0.279	0.433	0.213	0.269	45.7
B.5.1	100C1	0.80	0.86	3.25	2.40	0.03	0.274	0.274	0.390	0.213	0.285	48.4
B.5.3	100C1	0.80	0.86	3.25	2.00	0.03	0.274	0.274	0.390	0.202	0.323	54.9
B.5.4	100C1	0.70	0.90	3.10	2.40	0.03	0.287	0.287	0.418	0.200	0.285	48.4
B.5.7	100C1	0.60	0.98	3.00	2.40	0.03	0.312	0.312	0.446	0.183	0.285	48.4
B.5.9	100C1	0.60	0.98	3.00	2.60	0.03	0.312	0.312	0.446	0.187	0.269	45.7
B.5.10	100C1	0.70	0.90	3.10	2.60	0.03	0.287	0.287	0.418	0.205	0.269	45.7
B.6.1	70C1+30PC	0.80	0.77	2.35	2.40	0.03	0.350	0.350	0.405	0.222	0.285	48.4
B.6.2A	70C1+30PC	0.80	0.77	2.35	2.20	0.03	0.350	0.350	0.405	0.216	0.303	51.5
B.6.2B	70C1+30PC	0.80	0.77	2.35	2.20	0.03	0.350	0.350	0.405	0.216	0.303	51.5
B.6.3A	70C1+30PC	0.80	0.77	2.35	2.00	0.03	0.350	0.350	0.405	0.209	0.323	54.9
B.6.3B	70C1+30PC	0.80	0.77	2.35	2.00	0.03	0.350	0.350	0.405	0.209	0.323	54.9
B.6.4	70C1+30PC	0.70	0.81	2.30	2.40	0.03	0.369	0.369	0.432	0.207	0.285	48.4
B.6.5S	70C1+30PC	0.70	0.81	2.30	2.20	0.03	0.369	0.369	0.432	0.202	0.303	51.5
B.6.7S	70C1+30PC	0.60	0.89	2.20	2.40	0.03	0.405	0.405	0.460	0.189	0.285	48.4

B.6.9	70C1+30FC	0.60	0.89	2.20	2.60	0.03	0.405	0.405	0.460	0.194	0.269	45.7
B.6.10	70C1+30FC	0.70	0.81	2.30	2.60	0.03	0.369	0.369	0.432	0.212	0.269	45.7
B.6.10S	70C1+30FC	0.70	0.81	2.30	2.60	0.03	0.369	0.369	0.432	0.212	0.269	45.7
B.7.1	70C1+30FG	0.80	0.85	2.50	2.40	0.03	0.387	0.387	0.394	0.216	0.285	48.4
B.7.10	70C1+30FG	0.70	0.89	2.50	2.60	0.03	0.405	0.405	0.421	0.207	0.269	45.7
B.8.1	70C1+30CV	0.80	0.75	2.50	2.40	0.03	0.341	0.258	0.407	0.223	0.285	48.4
B.8.2	70C1+30CV	0.80	0.75	2.50	2.20	0.03	0.341	0.258	0.407	0.217	0.303	51.5
B.8.3	70C1+30CV	0.80	0.75	2.30	2.00	0.03	0.341	0.258	0.407	0.211	0.323	54.9
B.8.4	70C1+30CV	0.70	0.79	2.30	2.40	0.03	0.359	0.271	0.434	0.208	0.285	48.4
B.8.9	70C1+30CV	0.60	0.86	2.45	2.60	0.03	0.391	0.295	0.464	0.195	0.269	45.7
B.8.10	70C1+30CV	0.70	0.79	2.50	2.60	0.03	0.359	0.271	0.434	0.213	0.269	45.7
B.12.1	50C1+50FC	0.80	0.71	1.00	2.40	0.03	0.452	0.452	0.418	0.229	0.285	48.4
B.12.10	50C1+50FC	0.70	0.75	1.00	2.60	0.03	0.478	0.478	0.446	0.218	0.269	45.7

MATERIALS AND METHODS

B.13.1	40C1 + 60PC	0.80	0.70	0.80	2.40	0.03	0.557	0.557	0.421	0.230	0.285	48.4
B.13.10	40C1 + 60PC	0.70	0.74	0.80	2.60	0.03	0.589	0.589	0.448	0.220	0.269	45.7
B.14.1	60C2 + 40PC	0.80	0.69	1.30	2.40	0.03	0.378	0.378	0.420	0.230	0.285	48.4
B.14.10	60C2 + 40PC	0.70	0.73	1.35	2.60	0.03	0.400	0.400	0.447	0.219	0.269	45.7
B.15.1	50C2 + 50PC	0.80	0.68	1.15	2.40	0.03	0.447	0.447	0.422	0.231	0.285	48.4
B.15.10	50C2 + 50PC	0.70	0.72	1.15	2.60	0.03	0.474	0.474	0.449	0.220	0.269	45.7

Table 2: Effective dosages of SCC per cubic metre

Id.	Effective dosages in kg for solids and in litres for water and SP									
	C1	C2	FC	CV	FG	S1	S2	G1	G2	W	SP
B.2.1		543	121			289	437	387	385	170	13.3
B.2.2		528	118			281	425	411	409	165	12.9
B.2.3		512	115			273	412	438	437	160	12.5
B.2.4		505	113			307	464	387	385	168	12.4
B.2.5		492	110			299	452	411	409	164	12.0
B.2.6a		477	107			290	439	438	437	159	11.7
B.2.6b		477	107			290	439	438	437	159	11.7
B.2.7		459	103			326	493	387	385	170	11.0
B.2.8		447	100			318	480	411	409	166	10.7
B.2.9		470	105			334	504	365	364	174	11.2
B.2.10		517	116			314	475	365	364	172	12.6
B.3.1		532			116	283	428	387	385	179	13.3
B.3.2		518			113	276	417	411	409	175	12.9
B.3.3		502			110	268	404	438	437	169	12.5
B.3.4		496			108	302	456	387	385	177	12.4
B.3.5		483			105	294	444	411	409	173	12.0
B.3.6		468			102	285	431	438	437	168	11.7
B.3.7		453			99	322	486	387	385	177	11.0
B.3.9		464			101	329	497	365	364	181	11.3
B.3.10		507			111	309	467	365	364	181	12.7
B.4.1		543		106		289	437	387	385	170	13.0
B.4.3		513		100		273	413	438	437	160	12.3
B.4.4		505		99		307	465	387	385	168	12.1
B.4.5		492		96		299	453	411	409	164	11.8
B.4.7		461		90		327	495	387	385	169	10.8
B.4.9		472		92		335	506	365	364	173	11.0
B.4.10		517		101		315	475	365	364	172	12.4
B.5.1	670					277	418	387	385	184	20.8
B.5.3	633					261	395	438	437	173	19.6
B.5.4	628					296	448	387	385	180	18.6
B.5.7	575					316	478	387	385	180	16.4
B.5.9	589					324	489	365	364	184	16.8

	C1	C2	C3	C4	C5	C6	C7	C8	C9	C10	C11
B.5.10	643					303	458	365	364	184	19.0
B.6.1	488		181			287	434	387	385	171	15.0
B.6.2a	475		176			280	423	411	409	176	14.6
B.6.2b	475		176			280	423	411	409	176	14.6
B.6.3a	460		171			271	410	438	437	161	14.1
B.6.3b	460		171			271	410	438	437	161	14.1
B.6.4	455		169			307	464	387	385	168	13.7
B.6.5s	444		165			299	451	411	409	165	13.3
B.6.7s	416		154			327	494	387	385	154	12.0
B.6.9	425		158			334	505	365	364	172	12.2
B.6.10	466		173			314	474	365	364	172	14.0
B.6.10s	466		173			314	474	365	364	172	14.0
B.7.1	475				172	280	423	387	385	184	15.4
B.7.10	454				164	306	462	365	364	184	14.7
B.8.1	490			159		289	437	387	385	167	15.5
B.8.2	478			155		281	425	411	409	163	15.1
B.8.3	463			150		273	412	438	437	158	14.6
B.8.4	458			149		308	466	387	385	165	14.4
B.8.9	428			139		337	509	365	364	168	13.2
B.8.10	468			152		315	477	365	364	168	14.8
B.12.1	360		312			297	449	387	385	163	6.4
B.12.10	343		297			323	489	365	364	164	6.1
B.13.1	289		376			298	451	387	385	161	5.1
B.13.10	276		358			325	491	365	364	163	4.8
B.14.1		420	250			298	450	387	385	159	8.3
B.14.10		400	239			324	490	365	364	160	8.2
B.15.1		351	314			299	453	387	385	157	7.3
B.15.10		335	300			326	493	365	364	159	7.0

Test Program

Mixing Procedure

The procedure of mixing the mortar was similar to that proposed by Domone and Jin [38], exception for a stop introduced to clean the blades of the mixer, as described on Ref. [26]. The procedure of the concrete mixing was similar to that adopted for mortar, excluding the stop for cleaning blades, which was no longer required here. All the components and the total water (W) were introduced into the mixer and mixing proceeded to the normal speed for six minutes, followed by a stop for two minutes, and again, a one-minute mixing at normal speed. The superplasticizer was added to the mixture one minute after its start and without stopping the mixer.

Fresh Properties

The tests and procedures to evaluate the rheological properties of the mortar phase and the experimental and iterative process used to obtain the adequate values of Vw/Vp and Sp/p% that provided the required rheological properties were described in Ref. [26]. The adequate values were those that fulfilled the flow requirements expressed as the relative spread area (Gm) and relative flow velocity (Rm). The admissible interval for Gm was between 5.3 and 5.9, corresponding to a slump-flow diameter (Dm) between 251 and 263 mm. The admissible interval for Rm was between 1.14 and 1.30 s−1, which corresponds to a v-funnel flow time (t) between 7.69 and 8.77 s.

The tests for the evaluation of the fresh properties of concrete were divided in two main parts, namely, the indirect evaluation of the rheological properties and the evaluation of self-compactability properties. The achievement of adequate rheological properties of concrete was considered a necessary condition, but not a sufficient one, to ensure the self-compactability. Accordingly, concretes that fulfil the required rheological properties are not necessarily self-compactable concretes. Additionally, the evaluation of the flow properties of concrete has also the purpose of test the adequacy of the flow properties of mortar previously defined.

The rheological properties were evaluated indirectly by the slump-flow and v-funnel tests. The test equipment and the experimental procedure for slump-flow test was the one described by RILEM TC 174-SCC [39], which is similar to NP EN 12350-8:2010 [40]. Also, the test equipment and experimental procedure for v-funnel test was that described by RILEM TC 174-SCC [41], which is similar to NP EN 12350-9:2010 [42]. All the concretes should have a relative spread area (Gc) between 8 and 11.25, corresponding to slump-flow diameter (Dm) from 600 to 700 mm and, simultaneously, a relative flow velocity (Rc) between 0.5 and 1.0 s−1, corresponding to v-funnel time (t) from 10 to 20 s. The relative spread area (Gc) and the relative flow velocity (Rc) are given by Eqs. (1) and (2), respectively, where (Dm) stands for the average spread diameter in mm, (D0) stands for the initial diameter at the base of the cone in mm and (t) stands for the v-funnel time in seconds.

$$Gc = \left(\frac{Dm}{D_0}\right)^2 - 1 \quad (1)$$

$$Rc = \frac{10}{t} \quad (2)$$

The evaluation of self-compactability was accessed by the L-box test and Box test. These tests implicitly evaluate the rheological properties but, at the same time, reflect other workability properties of SCC, including the filling ability, resistance to segregation and the passing ability. Box-test device appeared as a modification of the U-box test with a more severe flowing resistance due to the angular shape on the base. The measured parameters in the L-box test were the passing ratio H2/H1 and the T40 (time that the concrete front takes to reach 400 mm apart from obstacles), using the test equipment and experimental procedure described by RILEM TC 174-SCC [43], which is similar to NP EN 12350-10:2010 [44]. In the Box test, only the filling height (H) was measured using the test equipment and experimental procedure of RILEM TC 174-SCC [45].

Hardened Properties

For the mortar mixtures that met the flow requirements, 4 cubic specimens of 50 mm side were moulded. After moulding, all samples were protected by plastic sheeting to prevent premature loss of water during 24 h and then cured in water at 20 ± 2 °C. At 28-days age, saturated surface dry density and the compressive strength was measured. For each one of the 60 produced concretes, 6 cubic specimens of 150 mm side were moulded, protected by plastic sheeting during 24 h and then cured in water at 20 ± 2 °C. At 7-days age, compressive strength (fc,7) was measured and at 28-days age, saturated surface dry density (δc,28) and compressive strength (fc,28) were measured.

TEST RESULTS AND DISCUSSION

Fresh Properties

The study of the mortar phase was described in Ref. [26]. The most significant results obtained in mortars with binary blends of powders under identical flow properties are summarised here. The particle size distribution of fine aggregates remained constant and was represented by the proposed reference curve. It was shown that a close coordination between the proposed mix design parameters (mixture of powders, Vp/Vs, Vw/Vp and Sp/p%) is needed to obtain simultaneously the required workability and compressive strength. To ensure the desired rheological properties, all parameters of the mixture are crucial. However, to ensure the attainment of a particular W/C ratio, which leads to a certain compressive strength, the variation on the SP dosage (Sp/p%) is less important. The viscosity is controlled primarily by the dosage of powders and by W/C ratio, while the yield stress is controlled mainly by the SP. If in a given mixture the SP is replaced by another, the Vw/Vp ratio will not vary significantly. This means that the proposed methodology is not restricted to the type of SP studied. For this reason, the ratios Vw/Vp and Sp/p% have to be experimentally confirmed to assess the correct dosage of SP. In last years the authors have found reductions of about 30–50% for the same polycarboxylate-based superplasticizer as it was improved by the suppliers, and this has showed not to produce any change in the methodology used, since the SP dosage is determined experimental in the mortar phase. Additionally, in binary mixes, it was found that the correlation between the W/C ratio and the percentage of cement replacement by the addition depends on the Vp/vs. This fact led to the conclusion that, once the powder materials (cement and addition) have been selected and the Vp/Vs has been decided, it would be feasible to estimate the percentage of cement to be replaced by the addition that leads to a specified W/C ratio. This has enabled the definition of a simple methodology for the mix design of the mortar phase in binary blends of powder to obtain both the adequate rheological properties and the required W/C ratio.

The study on concrete phase is discussed in this paper. For convenience, fresh and hardened properties of concretes are presented in Table 3. The hardened properties will be discussed on Section 3.2. Concerning to the fresh properties, Figure 2 shows the values of Gc and Rc obtained in all mixtures. From the analysis of Figure 2 it was concluded that the rheological properties defined for the mortar phase were suitable for obtaining the desired rheological properties in concrete phase. For the observed correspondence, the absence of segregation in the slump-flow test and the absence of blocking in the v-funnel test in all mixtures had greatly contributed. Segregation was evaluated visually by observing concrete behaviour in slump-flow, v-funnel and L-box test. The results showed that the correlation between the rheological properties of mortars and concretes needed be evaluated under similar test conditions, i.e., by their respective slump-flow and v-funnel tests, and in the absence of segregation and blocking. The correspondence between the rheological properties of mortars and concretes occurred regardless of the values of Vp/Vs and Vm/Vg. This means that the fresh concrete fulfilled the first functional requirement relating to the rheology, less restrictive, but nevertheless essential. The self-compactability requirements, considered to be more restrictive, were the second functional requirement and, in this case, Vp/Vs and Vm/Vg had a more significant importance.

Table 3: Fresh and hardened properties of SCC.

Id.	MN	Fresh properties of concrete											Hardened properties of concrete			
		Slump-flow				V-funnel		L-box		Box	Density	Compressive strength	
		T50 (s)	Dm (mm)	Gc	t (s)	Rc (s⁻¹)	T40 (s)	H2/H1	H (mm)	δc,28 (kg/m³)	fc,7 (MPa)	fc,28 (MPa)
B.2.1	1.92	1.35	695	11.08	11.27	0.89	3.22	0.90	327	2384	49.5(0.7)	63.7(0.4)
B.2.2	1.76	2.01	670	10.22	11.14	0.90	4.96	0.84	325	2430	49.8(0.9)	66.3(0.1)
B.2.3	1.60	2.36	625	8.77	10.39	0.96	5.16	0.79	323	2406	50.6(0.4)	64.8(0.3)
B.2.4	1.68	2.63	620	8.61	11.84	0.84	5.69	0.81	323	2374	45.7(0.8)	60.4(0.5)
B.2.5	1.54	2.71	630	8.92	13.15	0.76	6.90	0.78	323	2418	46.2(0.4)	60.0(1.4)
B.2.6A	1.40	4.31	600	8.00	18.53	0.54	8.50	0.64	316	2416	47.2(0.5)	59.9(1.2)
B.2.6B	1.40	3.60	630	8.92	14.95	0.67	8.56	0.71	320	2418	46.6(0.5)	58.0(0.6)
B.2.7	1.44	2.81	645	9.40	10.25	0.98	5.29	0.73	318	2360	40.1(0.9)	50.5(0.7)
B.2.8	1.32	3.61	625	8.77	13.03	0.77	8.08	0.62	315	2402	38.7(1.0)	49.9(0.6)
B.2.9	1.56	2.71	610	8.30	9.91	1.01	4.56	0.74	318	2384	38.2(2.1)	50.9(1.7)
B.2.10	1.82	3.17	645	9.40	9.15	1.09	4.52	0.86	325	2402	46.1(0.1)	60.3(0.7)
B.3.1	1.92	2.72	660	9.89	10.96	0.91	4.59	0.85	329	2394	47.6(2.0)	59.7(1.1)
B.3.2	1.76	3.64	635	9.08	14.03	0.71	5.05	0.81	323	2414	47.8(1.1)	59.7(1.0)
B.3.3	1.60	3.41	630	8.92	15.49	0.65	7.63	0.77	320	2420	47.8(1.5)	58.5(1.1)

TEST RESULTS AND DISCUSSION

B.3.4	1.68	2.55	630	8.92	11.45	0.87	5.18	0.81	325	2386	41.7(1.4)	53.0(0.3)
B.3.5	1.54	3.56	655	9.73	12.47	0.80	5.27	0.81	322	2388	43.7(1.0)	52.9(0.5)
B.3.6	1.40	2.95	635	9.08	16.13	0.62	5.85	0.73	320	2398	43.8(0.8)	53.3(0.5)
B.3.7	1.44	2.76	615	8.46	11.27	0.89	6.49	0.67	317	2394	36.5(2.1)	45.7(1.2)
B.3.9	1.56	2.86	605	8.15	10.06	0.99	5.16	0.73	319	2357	35.8(1.1)	44.6(1.0)
B.3.10	1.82	2.76	640	9.24	10.12	0.99	4.03	0.85	323	2382	42.7(0.2)	54.3(1.0)
B.4.1	1.92	2.25	650	9.56	9.98	1.00	3.93	0.84	323	2394	54.1(1.3)	69.1(0.7)
B.4.3	1.60	3.12	615	8.46	12.65	0.79	5.52	0.77	320	2406	55.9(0.4)	67.5(1.3)
B.4.4	1.68	2.31	625	8.77	9.73	1.03	4.06	0.80	319	2382	49.6(0.3)	63.1(1.0)
B.4.5	1.54	3.06	635	9.08	10.97	0.91	4.97	0.77	319	2400	50.4(0.7)	64.8(1.0)
B.4.7	1.44	2.43	625	8.77	11.59	0.86	5.52	0.70	315	2382	44.1(0.2)	59.6(1.5)
B.4.9	1.56	2.55	630	8.92	10.29	0.97	4.09	0.77	322	2376	43.0(0.9)	56.8(0.3)
B.4.10	1.82	3.28	640	9.24	11.62	0.86	4.23	0.83	324	2396	50.4(0.4)	63.6(1.1)
B.5.1	1.92	3.13	620	8.61	10.38	0.96	4.74	0.81	320	2404	74.5(2.2)	85.0(2.4)
B.5.3	1.60	3.40	635	9.08	13.24	0.76	5.26	0.79	324	2416	76.2(0.1)	86.8(1.8)

B.5.4	1.68	3.54	615	8.46	11.57	0.86	5.27	0.77	321	2400	70.8(2.6)	82.9(1.0)
B.5.7	1.44	2.97	645	9.40	13.25	0.75	5.24	0.75	317	2406	67.7(0.5)	77.8(1.0)
B.5.9	1.56	3.25	640	9.24	10.63	0.94	4.39	0.75	317	2384	65.3(0.9)	76.8(0.1)
B.5.10	1.82	3.82	630	8.92	9.94	1.01	4.89	0.86	325	2390	72.3(2.5)	81.9(1.0)
B.6.1	1.92	3.08	620	8.61	12.51	0.80	4.91	0.81	322	2416	66.8(0.3)	78.4(0.5)
B.6.2A	1.76	3.24	635	9.08	12.96	0.77	4.99	0.83	325	2414	69.9(1.0)	77.4(1.6)
B.6.2B	1.76	3.34	645	9.40	11.67	0.86	4.36	0.84	325	2416	68.1(0.4)	79.2(0.6)
B.6.3A	1.60	3.70	675	10.39	11.04	0.91	5.24	0.88	328	2436	69.2(0.7)	78.7(1.4)
B.6.3B	1.60	4.05	645	9.40	14.05	0.71	5.51	0.83	325	2430	67.0(1.2)	77.9(1.4)
B.6.4	1.68	3.29	630	8.92	14.93	0.67	6.71	0.80	323	2402	63.6(3.3)	74.7(2.1)
B.6.5S	1.54	4.02	685	10.73	16.52	0.61	6.91	0.89	328	2445	68.9(0.8)	79.3(1.1)
B.6.7S	1.44	4.06	680	10.56	14.76	0.68	6.10	0.80	324	2428	59.7(0.4)	70.7(1.7)
B.6.9	1.56	3.45	650	9.56	12.70	0.79	5.31	0.83	325	2396	57.2(0.3)	66.8(1.3)
B.6.10	1.82	3.43	640	9.24	11.25	0.89	3.69	0.84	324	2392	63.9(2.1)	72.1(1.0)
B.6.10S	1.82	3.80	700	11.25	10.06	0.99	4.32	0.93	329	2430	67.8(0.9)	77.8(0.8)
B.7.1	1.92	3.22	630	8.92	12.52	0.80	5.03	0.82	324	2390	63.1(0.9)	74.1(0.3)
B.7.10	1.82	3.27	670	10.22	11.38	0.88	4.23	0.92	327	2374	59.4(0.1)	70.5(0.9)
B.8.1	1.92	4.07	655	9.73	10.51	0.95	4.42	0.84	327	2388	65.4(0.1)	80.3(1.2)
B.8.2	1.76	3.77	630	8.92	12.35	0.81	5.30	0.79	322	2378	62.8(0.4)	79.2(1.0)
B.8.3	1.60	3.80	645	9.40	13.30	0.75	6.17	0.80	325	2408	63.9(2.5)	78.2(2.8)
B.8.4	1.68	3.28	640	9.24	10.82	0.92	5.23	0.83	326	2384	62.2(1.7)	74.5(0.3)
B.8.9	1.56	3.34	655	9.73	11.33	0.88	3.98	0.82	326	2374	57.3(0.9)	69.9(1.6)

TEST RESULTS AND DISCUSSION

B.8.10	1.82	3.20	635	9.08	10.09	0.99	5.65	0.83	327	2388	62.5(1.0)	75.7(0.4)
B.12.1	1.92	4.20	620	8.61	14.93	0.67	5.82	0.77	319	2416	60.5(0.5)	69.1(0.7)
B.12.10	1.82	4.31	650	9.56	12.61	0.79	5.40	0.80	322	2398	54.1(0.4)	62.8(0.1)
B.13.1	1.92	3.40	660	9.89	10.99	0.91	4.20	0.85	324	2394	47.4(1.5)	54.3(1.1)
B.13.10	1.82	3.21	655	9.73	11.99	0.83	4.94	0.85	326	2394	42.7(0.5)	51.3(0.4)
B.14.1	1.92	3.30	640	9.24	12.18	0.82	6.40	0.86	322	2392	40.9(0.6)	50.9(1.1)
B.14.10	1.82	3.36	655	9.73	11.79	0.85	5.07	0.85	324	2376	37.6(0.1)	46.7(0.8)
B.15.1	1.92	3.46	670	10.22	10.41	0.96	5.38	0.88	327	2370	33.5(0.2)	42.5(0.5)
B.15.10	1.82	3.76	650	9.56	11.27	0.89	4.79	0.86	323	2357	29.6(0.4)	38.3(0.8)

The number between brackets signify the corresponding standard deviation.

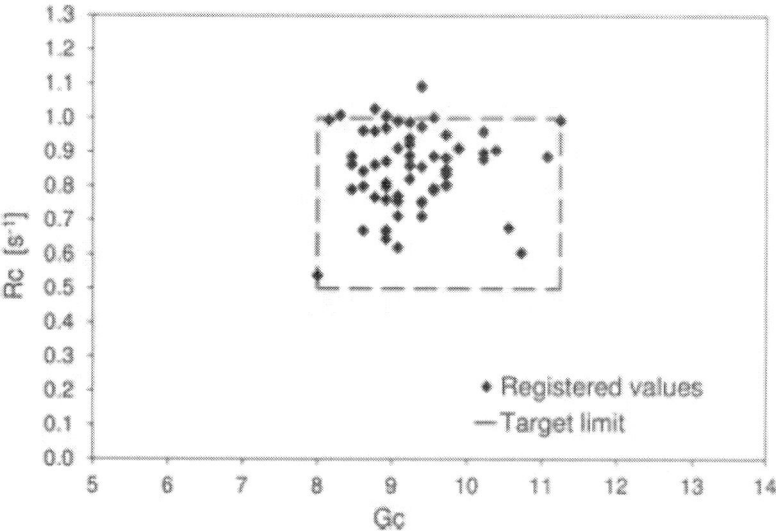

Figure 2: Target limit and registered values of Gc and Rc.

In the evaluation of SCC in the fresh state, the T50 (time that flow takes to reach the 500 mm mark in the slump-flow test) and T40 (in L-box) parameters have been suggested as an alternative to the measurement of v-funnel flow time (t). The relationship between the parameters t, T50 and T40, or between them and the plastic viscosity of concrete, is complex. It has been reported that the parameters t, T50 and T40 are affected by the slump-flow and, therefore, should only be correlated to the viscosity under a constant slump-flow [46]. Moreover, the measurement of the parameters t, T50 or T40 is performed in different devices with different restriction to flow, which may lead to differences in the interaction between the aggregate particles and influence the correlations between these parameters. Figure 3 shows the relationship between the flow times measured in the v-funnel test (t) and in the slump-flow test (T50), whileFigure 4 shows the relationship between the flow times measured in the v-funnel test (t) and in the L-box test (T40). The results have shown a very weak correlation between the parameters t and T50, and some improvement when parameter t was correlated with the T40. A first explanation for the weak correlation coefficient between the analysed parameters was attributed to the fact thatcorrelations had included mixtures with different amounts of coarse aggregates, leading to different degrees of interaction between particles in different forms of testing. It was concluded that the T50 and T40 parameters should not replace the v-funnel test in the mix design stage. Eventually, for the quality control of production on site, the T50 or T40 parameters may replaces the v-funnel test to detect variations in the mix proportions.

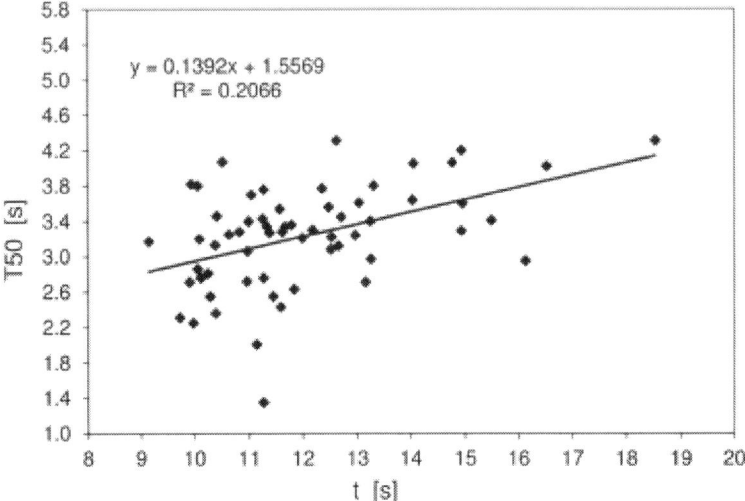

Figure 3: Flow time t in v-funnel versus T50 in slump-flow.

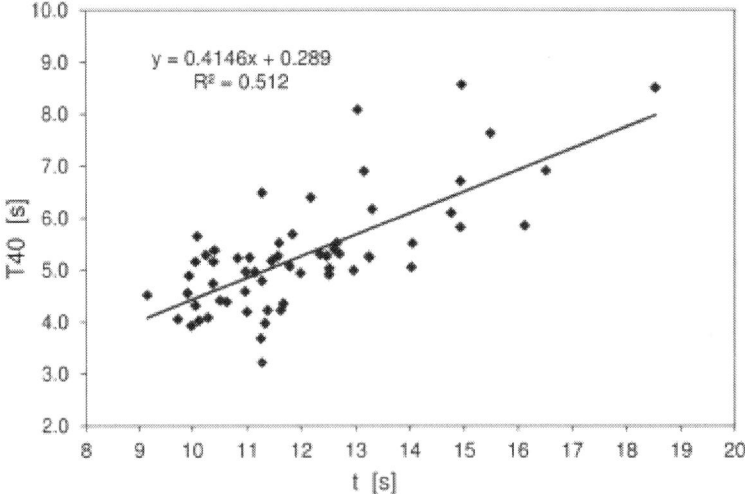

Figure 4: Flow time t in v-funnel versus T40 in L-box.

The L-box test and Box test were used in the analysis of the functional requirement of self-compactability and the corresponding parameters H2/H1 and H. In both trials, no blocking or visible segregation phenomena were observed for all the mixtures. Under these conditions, it can be assumed that different concrete mixtures, with different rheological parameters within the target set, may lead to the same self-compactability parameter, either in the L-Box test or in the Box test. Consequently, self-compactability parameters cannot be analysed independently of the rheological properties. In the present study it was observed that the parameters of self-compactability were more affected by the flow spread diameter (Dm), while the variation of flow time in v-funnel test (t) had a minor effect. The lowest influence of the v-funnel time was probably due to the fact that proper viscosity was ensured by the range of the volume of powders and the W/C ratios used.

Figure 5 shows the variation of H2/H1 with Dm, while Figure 6 shows the variations of H with Dm. It was observed that, in the absence of blocking, the descent of the concrete in the L-box test and the filling height in the Box test strongly depends on the flowability of concrete (slump-flow), and hence, self-compactability (passing ability) could only be compared between mixtures with the same slump-flow. For higher values of MN, the correlation between Dm and H has showed less sensitivity on the evaluation of the different levels of self-compactability (passing ability), compared to the correlation with H2/H1 parameter. To that extent, it may be assumed that, after a certain amount of reduction of aggregates, the Box test becomes less sensitive compared to the L-box test. To evaluate the influence of the mix design parameters in the self-compactability properties it was then necessary to isolate the resulting influence of rheological properties. In view of the adopted methodology, the only way to achieve this goal was by establishing a relationship between mix design parameters, the rheological parameters (filling ability) and the self-compactability parameters (passing ability). This analysis was performed separately for the L-box and Box tests.

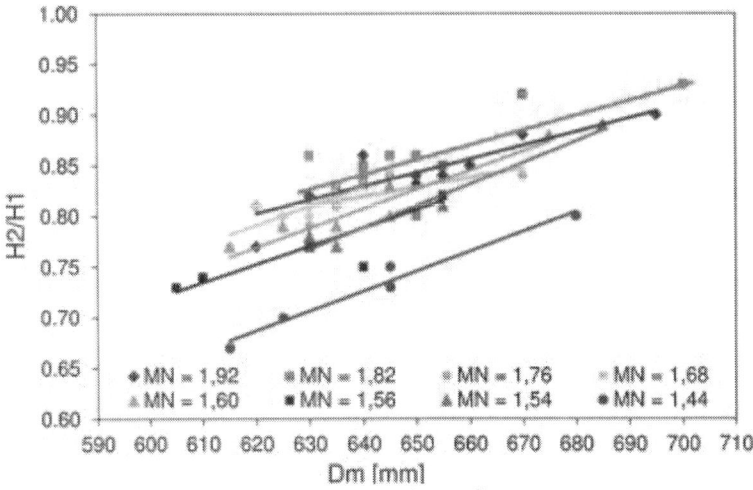

Figure 5: H2/H1 in L-box test versus Dm in slump-flow test.

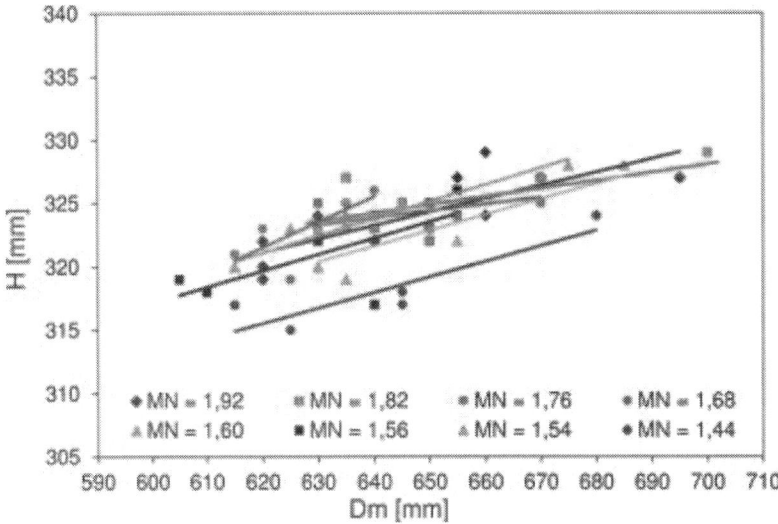

Figure 6: H in Box test versus Dm in slump-flow test.

In the L-box test, the parameters used for analysis were H2/H1, Dm and MN. To study the combined effects of these variables, multi-variable regression analysis has been carried out to derive the best-fit curve. The best-fit curve is shown graphically alongside the test points in Figure 7 and is expressed by Eq. (3). In Eq. (3), the constants *a*, *b* and *c* have assumed the following values: *a* = 2.189, *b* = −0.0997 and *c* = −673.283. A high R2 value of 0.90 has been achieved, indicating that the H2/H1 is highly related to MN and Dm.

$$\left(\frac{H2}{H1}\right) = a + b \times \frac{MN}{\ln(MN)} + \frac{c}{Dm} \qquad (3)$$

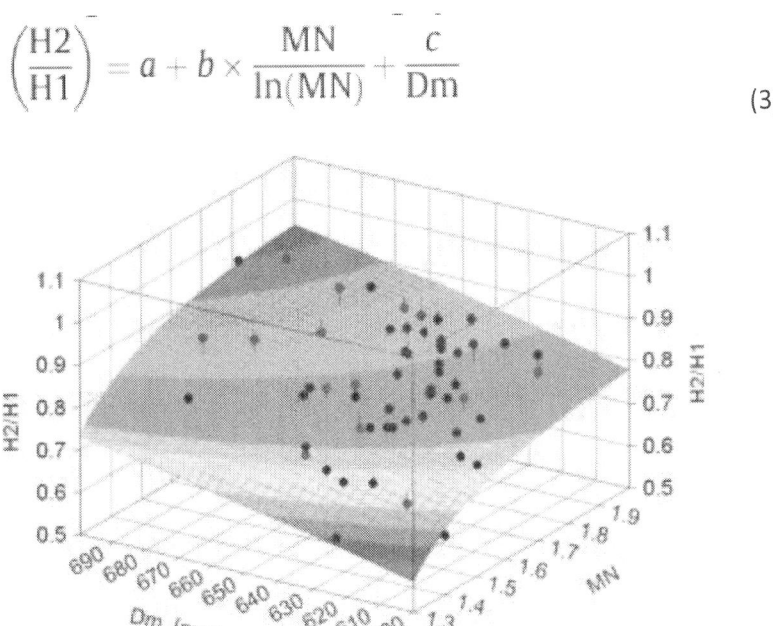

Figure 7: Best-fit curve for H2/H1 as a function of Dm and MN.

In the Box test, the parameters used for analysis were the H, Dm and MN. The best-fit curve is shown graphically alongside the test points in Figure 8 and is expressed by Eq. (4). In Eq. (4), the constants a, b and c have assumed the following values: $a = 370.4$, $b = -4.07$ and $c = -2160665$. A much lower R2 value of 0.68 was achieved in comparison to that obtained for the L-box test, most likely due to the difficulty in reflecting the loss of sensitivity previous mentioned.

$$H = a + b \times \frac{(MN)}{\ln(MN)} + c \times \left(\frac{\ln(Dm)}{Dm^2}\right)$$

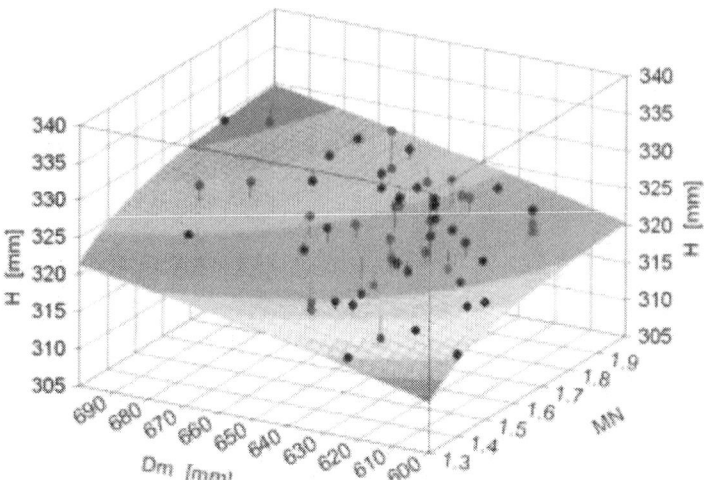

Figure 8: Best-fit curve for H as a function of Dm and MN.

The MN represents the product of Vp/Vs by Vm/Vg and shown to be highly related with the total volume of aggregates (Figure 9). For constant values of Vp/Vs and Vm/Vg, the volume of the paste was almost constant. In that situation, changes in fresh properties of SCC were primarily due to changes in the composition of the paste. On the other hand, when using the same paste and different values of MN, the changes in the fresh properties of SCC occurred as a result of the simultaneous change of the volume of paste and proportions between the fine and coarse aggregates.

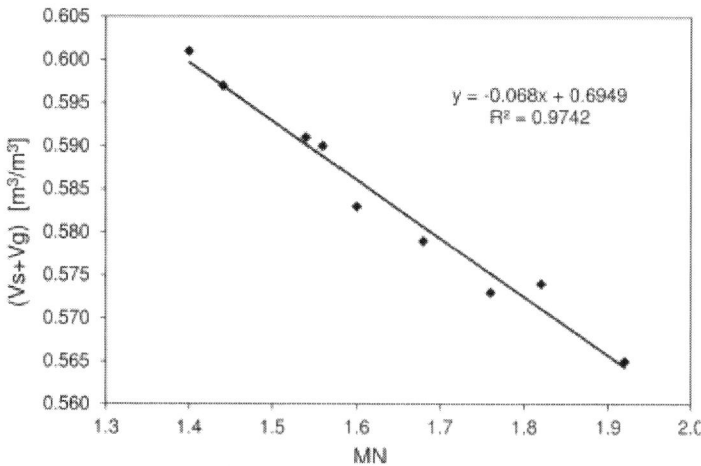

Figure 9: MN versus total volume of aggregates.

There is a broad consensus that a target filling height (H) in the Box or U-box tests higher or equal to 300 mm is adequate to ensure the adequate self-compactability. In the L-box test, concrete is usually considered self-compactable when H2/H1 is higher or equal to 0.80. It should be noted that all concretes produced in this study complied with the criteria set for the Box test, while only a part of them fulfilled the criteria set for the L-box test. Thus, it was concluded that the requirement for the L-box test (H2/H1 ⩾ 0.80) is more restrictive than that imposed for the Box test (H ⩾ 300 mm). In other words, a concrete can be considered self-compactable according to the criteria set for the Box test, and not be considered as self-compactable if the criteria for L-box test is used. In general, each of these test devices is associated with a particular procedure for the mix design of SCC and, therefore, has been used separately. Figure 10 shows the relationship between the H2/H1 and H values recorded during the experimental programme, which shows a probable correlation of linear type. If this correlation is established using the H2/H1 and the H determined based on Eqs. (3) and (4), respectively, it significantly improved (Figure 11). However, such a better correlation can hide the fact that the equation for the Box test has a low correlation coefficient. For the aforementioned reasons, the methodology for the mix design of SCC was mainly focused on the L-box test. The previously presented Eq. (3) can be written as a function of MN. Being an indeterminate equation, the software was again used to express MN as a function of Dm and H2/H1, based on the values that led to Eq.(3). This procedure led to Eq. (5) with the same R^2 of 0.90 and with the constants a, b, c, d, e, f and g being the following values: $a = 1.2214$, $b = -1.3605$, $c = 1.3099$, $d = -0.001758$, $e = -1.0184$, $f = 1.0855$ and $g = -0.001524$. Being MN the product of Vp/Vs by Vm/Vg, Eq. (5) can be expressed in order to Vm/Vg as shown in Eq. (6).

$$MN = \frac{a + b \times \left(\frac{H2}{H1}\right) + c \times \left(\frac{H2}{H1}\right)^2 + d \times (Dm)}{1 + e \times \left(\frac{H2}{H1}\right) + f \times \left(\frac{H2}{H1}\right)^2 + g \times (Dm)} \tag{5}$$

$$\left(\frac{Vm}{Vg}\right) = \frac{1}{\left(\frac{Vp}{Vs}\right)} \times \frac{a + b \times \left(\frac{H2}{H1}\right) + c \times \left(\frac{H2}{H1}\right)^2 + d \times (Dm)}{1 + e \times \left(\frac{H2}{H1}\right) + f \times \left(\frac{H2}{H1}\right)^2 + g \times (Dm)} \tag{6}$$

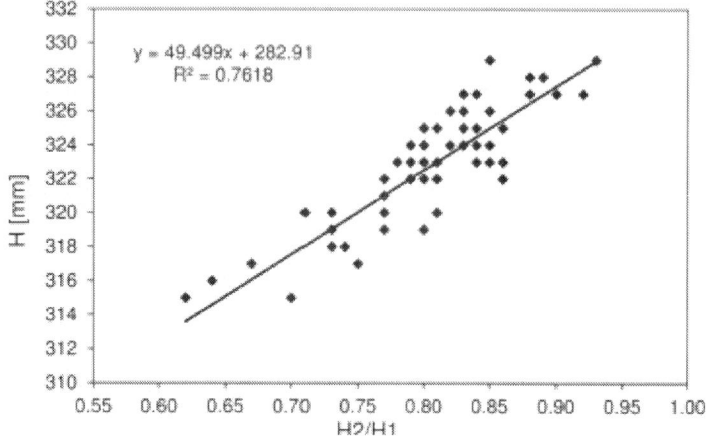
Figure 10: H2/H1 versus H based on registered values.

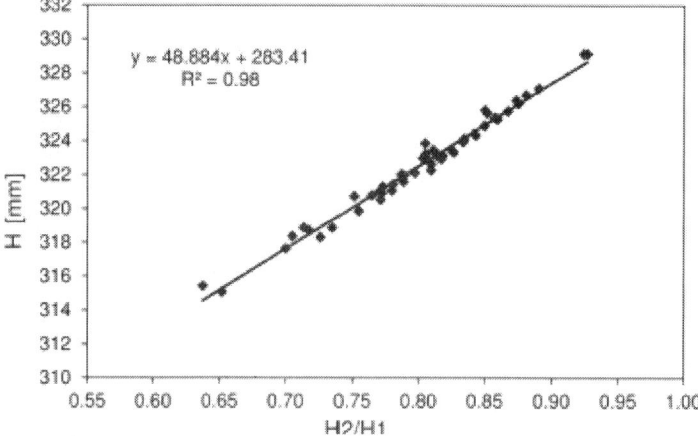
Figure 11: H2/H1 versus H based on best-fit curves.

Eq. (3) for the L-box test has enabled to isolate the effect of flowability (Dm) in the relationship between the mix design parameter (MN) and the self-compactability parameter (H2/H1). Using Eq. (3), it was found that under constant conditions of flowability (Dm), the self-compactability, represented by the parameter H2/H1 in L-box, depends on the combination between the absolute volume of fine aggregate in the mortar (Vs/Vm) and the absolute volume of coarse aggregates in concrete, being the last expressed by Vg, Vm/Vg or Vap. This behaviour is shown in Figure 12 for a constant Dm of 650 mm. It was observed that, when the volume of coarse aggregate in concrete mixtures increased, and consequently the volume of mortar decreased, compensation in the mortar phase was needed to keep the same level of self-compactability, by means of an increase in the ratio between the volume of paste and the volume of fine aggregate. Figure 13expresses the values of H2/H1 obtained by Eq. (3) as a function of equivalent parameters Vs/Vm and Vap used in general method, and assuming a constant Dm of 650 mm. When observed the pair of values 50% of Vap and Vs/Vm of 0.40 proposed in the general method for a successful SCC, it was concluded that more than a single value that checks the self-compactability parameter (H2/ H1 $\geqslant$ 0.80), it is possible to find a region of concretes that check this same parameter. Figure 14 shows the influence of the total volume of powders in the self-compactability (H2/H1) for a constant Dm of 650 mm. A convergence was observed between the correlations for the cement CEM II/B-L32.5N and the cement CEM I 42.5 R for a volume of powders of approximately 0.210 m3. It was also found that this limit value (Vp = 0.210 m3) is independent of flowability (Dm), since it happened for other values of Dm from 600 to 700 mm, not shown in the figure. This means that from a value of Vp of 0.210 m3, the self-compactability stops to depend on the type of cement and starts to depend solely on the total volume of powders. On the other hand, the volume of powder includes various binary combinations between cements and additions, and this appears to be less relevant to self-compactability than the total volume of powders for a Vp larger than 0.210 m3.

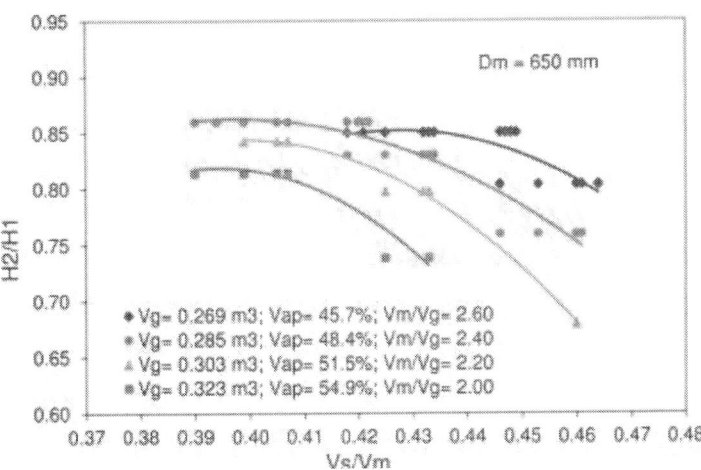

Figure 12: Vs/Vm versus H2/H1 for different Vm/Vg.

TEST RESULTS AND DISCUSSION

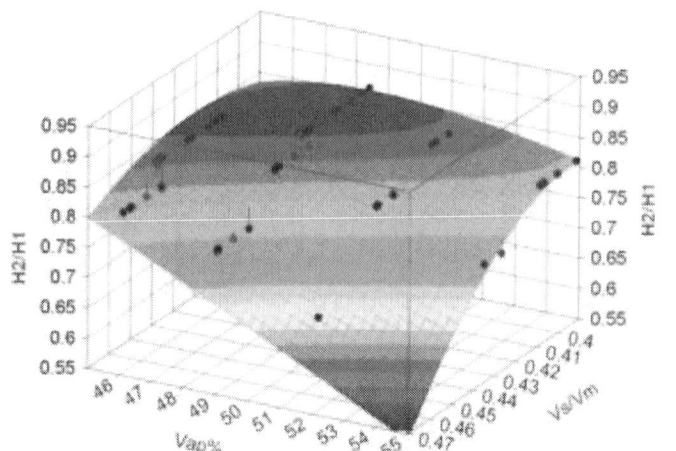

Figure 13: Comparison to the general method.

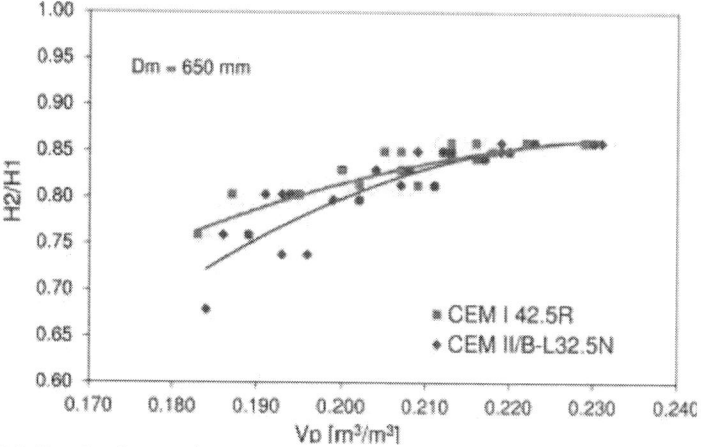

Figure 14: Total volume of powders versus H2/H1.

The binary mixtures with granite filler showed the highest average consumption of water, similar to the reference mixtures with cement only, while the binary mixtures with limestone powder and fly ash have shown lower and similar water consumptions. The dosage of the mixing water per cubic metre of SCC was in the range of 159 to 175 l when were used binary mixtures of cement (CEM I 42.5R or CEM II/B-L32.5N) with fly ash or limestone powder. The dosage of the mixing water per cubic metre of SCC was in the range of 169 to 187 l when were used cement-only mixtures or binary mixtures of cement (CEM I 42.5R or CEM II/B-L32.5N) with granite filler. The grading curve of the granite filler is close to that obtained for the fly ash and the specific surface fits the mean values obtained for the fly ash and limestone powder [26]. So, one can just speculate that the greatest need for water in mixtures with granite filler was due to particle shape. Vieira [47] have analysed images obtained by scanning electron microscopy (MEV) for granite filler of similar origin to that used in this work and found large dimensional variability and little regular particle shape. When cement pastes with additions of fly ash, limestone powder and granite filler were compared, the mixtures with granite filler has always showed the highest retained water ratio ßp, i.e., it requires more water to start flow [47]. The retained water ratio ßp, can be thought of as comprising the water adsorbed on the powder surface together with that required to fill the voids in the powder system and to provide sufficient dispersal of the particles for flow to be about to commence [48].

Hardened Properties

The discussion concerning the hardened properties of the mortar phase was presented in Ref. [26]. It was shown that the compressive strength of the mortar phase varied from 25 to 95 MPa and that exist a general correlation between the compressive strength of mortar and the W/C ratio (in mass) for each type of cement used, independent of the type and amount of the additions. Furthermore, it was observed a correlation between the compressive strength of mortar and Vp/Vs for each binary blend of powders. This study had shown that, for a certain cement and addition previously selected, it is possible to establish a relationship between the W/C ratio and the percentage of cement replacement by the addition. However, it was clear that such a relationship depends strongly on the Vp/Vs parameter defined for the mortar phase.

The results obtained in concrete phase are summarised in Table 3. The 28-days compressive strength of SCC was between 32 and 88 MPa. Figure 15 shows the correlations between the 28-days compressive strength of SCC and the W/C ratio (in mass) for the two types of cement used. As for the mortar phase, such correlations are independent of the type and amount of the additions. Due to the methodology used, the W/C ratio of the SCC mix is the same as its mortar phase, and this correlation can be used to find the W/C ratio that leads to a certain SCC compressive strength, and can also help on decision about the type of cement to be used. Correlations between the 28-days compressive strength of SCC and the water/ cementitious materials (W/CM) ratio showed slightly higher *R2* value to those obtained in the correlations with the W/C ratio. However, the correlations with the W/C ratio provide sufficient precision for initial studies of the mix design and can be more easily generalised, since it does not depend on the percentage of cement replacement by additions. For this reason, the correlations with the W/C ratio were taken into account in formulating the mix design methodology proposed in this paper, in contrast with the correlations with the W/CM ratio. The correlations between the compressive strength of SCC at 7 and 28 days age are presented in Figure 16, showing a marked difference in the early age strength between the two types of cement. Figure 17 shows the correlation between the 28-day compressive strength of mortars (fm,28) and concretes (fc,28). The concrete strength (fc,28) increased with the mortar strength (fm,28) at a gradually decreasing rate. For fm,28 up to about 75 MPa the concrete was stronger than its matrix. The opposite was observed for fm,28 higher than 75 MPa, showing the influence of the aggregates.

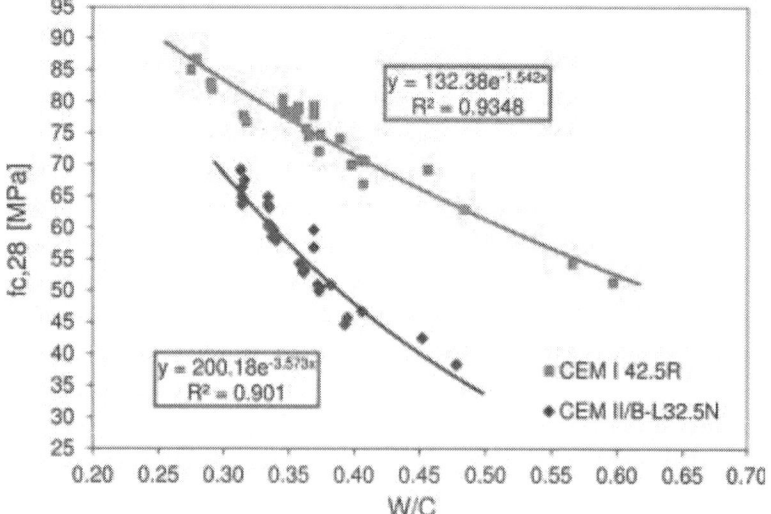

Figure 15: Compressive strength versus W/C ratio in mass.

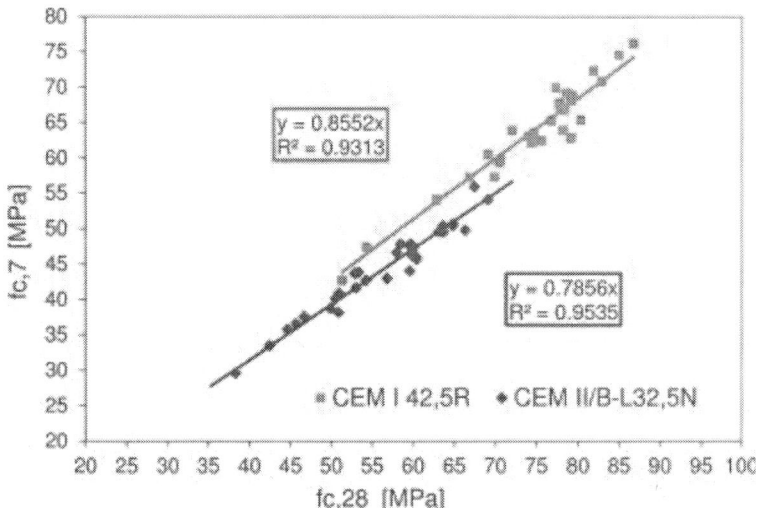
Figure 16: Compressive strength at 7 and 28 days age.

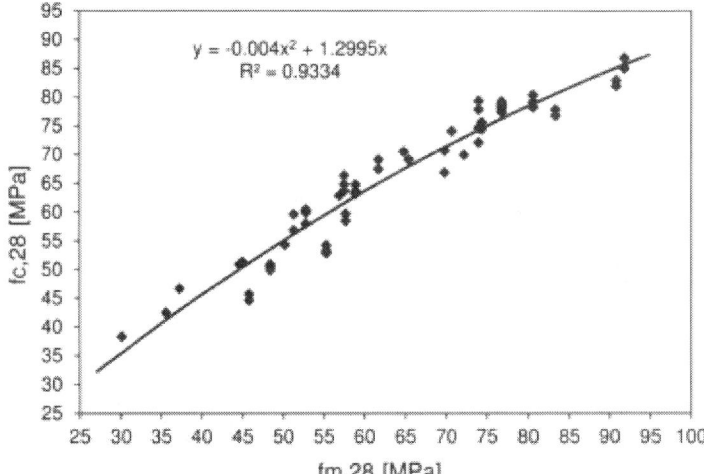
Figure 17: Compressive strength of mortars and concretes.

Proposed Methodology

The proposed methodology can be assumed as a step by step process, including the selection of the materials, studies in mortars and studies in concrete. The procedure considers the volumetric composition of the mix, with subsequent conversion to proportions by weight for batching.

The selection of materials starts on decision about the type of powder materials to be used. The type of cement could be selected taking into account the level of compressive strength to achieve on hardened concrete using the correlations presented in Figure 15, while the type of addition will depend mostly on its local availability. Preferably, a modified polycarboxylate based superplasticizer should be selected. The fine aggregates should comply with the reference grading curve proposed on Ref. [26], thus it is preferable to select two different fine aggregates with fineness modulus above and below that for the reference curve, to enable the determination of the proportions of such aggregates. The same procedure should be used for the coarse aggregates to achieve the reference curve proposed in Figure 1.

The studies in mortars imply a decision on the mix design parameters to be used, such as the combination of powder materials, Vp/Vs, Vw/Vp and Sp/p%. After have selected the type of cement and the level of strength, the W/C ratio is obtained by using Figure 15. The percentage of cement replacement by the selected addition is determined based on correlations presented in Ref. [26] as a function of the W/C ratio and the Vp/Vs value. For a start, a Vp/Vs value of 0.80 can be used. To decide on what values of Vw/Vp and Sp/p% to be used, general correlations were presented on Ref. [26], taking in mind that these values have to be adjusted by the experimental procedure described on Ref. [26] to obtain the adequate rheological properties of the mortar phase.

The studies on concrete phase imply the decision on the volume of voids (Vv) and the Vm/Vg. The Vv assumes a constant value of 0.03 m3/m3 (without air entraining admixtures) and the Vm/Vg is determined by Eq. (6) proposed in this paper as a function of the Vp/Vs, Dm and H2/H1.

Finally, trial mixtures can be produced and corrections of the mixture are admissible to fit all the requirements. The experience will dictate the best procedure. As a general rule, if only the water content (W) is increased, Rc and Gc values will increase simultaneously. Unlike, if only the Sp is increased, the Gc value will increase and the Rc value will experience a little or no increase. In both cases, increase of W or Sp, the H2/H1 value will probably increase, but caution has to be taken to avoid segregation. If the $f_{c,28}$ value is not achieved, changes in the powder mixture or in

Vp/Vs value have to be considered and all the process of mix design restarted.

CONCLUSIONS

A methodology for the first approach on mix design of SCC was presented. The methodology was based on simple procedures and assumes the SCC as a two phase material, the mortar phase and the coarse aggregates. Since the mortar phase properties were previously studied, the main achievement of this research was to evaluate the interaction between the coarse aggregates and the mortar phase and to define the maximum dosage of the coarse aggregates to be used.

Concerning the main achievement on this article, the following conclusions can be drawn:

- It was observed that the rheological properties defined for the mortar phase were suitable for obtaining the desired rheological properties in concrete phase. The results showed that the correlation between the rheological properties of mortars and concretes should be evaluated under similar test conditions, i.e., by their respective slump-flow and v-funnel tests, and in the absence of segregation in the slump-flow test and the absence of blocking in the v-funnel test.
- The results have shown a very weak correlation when comparing parameters t and T50. Better results were obtained when parameter t was correlated with the T40. It was clear that the T50 and T40 parameters should not replace the v-funnel flow time (t) in the mix design stage. Eventually, for the quality control of production on site, the T50 or T40 parameters might replaces the v-funnel test to detect variations in the mix proportions.
- Self-compactability parameters (H2/H1 or H) cannot be analysed independently from the rheological properties of concrete. It was observed that the parameters of self-compactability were more affected by the slump-flow (Dm), while the variation of v-funnel flow time (t) had a minor effect. The lowest influence of the v-funnel time was probably due to the fact that proper viscosity was ensured by the range of the volume of powders and the W/C ratios that were used.

- To evaluate self-compactability (passing ability), the Box test is less sensitive than the L-box test. The requirement for the L-box test (H2/H1 ⩾ 0.80) is more restrictive than that imposed for the Box test (H ⩾ 300 mm). A concrete can be considered self-compactable according to the criteria set for the Box test, and not be considered as self-compactable if the criteria for L-box test is used.
- Under constant conditions of flowability (Dm), the self-compactability, represented by the passing ability parameter H2/H1 in L-box, depends on the combination between the absolute volume of fine aggregate in the mortar (Vs/Vm) and the absolute volume of coarse aggregates in concrete (Vg). When Vg is increased, and consequently the volume of mortar decreases, compensation is needed in the mortar phase to keep the same level of self-compactability, by means of an increase in the ratio between the volume of paste and the volume of fine aggregate.
- A convergence was observed in the correlations between H2/H1 and the volume of powders (Vp) for the cements CEM II/B-L32.5N and CEM I 42.5 R, when Vp reached 0.210 m3. This limit value is independent of the flowability (Dm), since it happened for other values of Dm from 600 to 700 mm. This means that from a value of Vp of 0.210 m3, the self-compactability stops to depend on the type of cement and starts to depend solely on the total volume of powders. The volume of powder includes various binary blends of cements and additions, and this appears to be less relevant to self-compactability than the total volume of powders if Vp is higher than 0.210 m3.

REFERENCES

1. ACI 237 R-07 – Self-consolidating concrete. American Concrete Institute, USA, Reported by ACI committee 237, ISBN: 0-87031-244-8; 2007. 30p.
2. Bogas JA, Gomes A, Pereira MFC. Self-compacting lightweight concrete produced with expanded clay aggregate. Constr Build Mater 2012;35:1013–22.
3. Bogas JA. Characterization of structural lightweight expanded clay aggregate concrete (in Portuguese). PhD thesis in civil engineering. Portugal: Technical University of Lisbon; 2011.

4. Bui Van, Montgomery D. Mixture proportioning method for self-compacting high performance concrete with minimum paste volume. In: Proceedings of the 1st international RILEM symposium on SCC. Sweden; September 1999. p. 373–84.
5. Domone P, Hsi-Wen C. Testing of binders for high performance concrete. CemConcr Res 1997;27(8):1141–7.
1. 6.
6. Domone P. Mix design, self-compacting concrete: state-of-the-art report of RILEM technical committee 174-SCC, RILEM; 2000. p. 49–65.
7. Domone PL, Jin J, Chai HW. Optimum mix proportioning of self-compacting concrete. In: Proceeding on international conference on innovation in concrete structures: design and construction. London: University of Dundee, Thomas Telford; 1999. p. 277–85.
8. Domone PL. Self-compacting concrete: an analysis of 11 years of case studies. CemConcr Compos 2006;28(2):197–208.
9. Domone PLJ, Jin J. Properties of mortar for self-compacting concrete. In: Proceedings of the 1st international RILEM symposium on SCC. Sweden; 1999. p. 109–20.
10. Edamatsu Y, Nishida N, Ouchi M. A rational mix-design method for selfcompacting concrete considering interaction between coarse aggregate and mortar particles. In: Proceedings of the 1st international RILEM symposium on SCC. Sweden; September 1999. p. 309–20.
11. EFNARC. Specification and guidelines for self-compacting concrete. EFNARC, Norfolk, UK. ISBN: 0-9539733-4-4; February 2002. p. 1–32.
12. EN 206-1:2000. Concrete – Part 1: Specification, performance, production, and conformity, European Committee for Standardization; December 2000.
13. EPG (European Project Group), BIBM, CEMBUREAU, ERMCO, EFCA EFNARC. The European guidelines for self-compacting concrete: specification, production and use, EFNARC, UK; May 2005. p. 1–68.
14. Ho DWS, Sheinn AMM, Ng CC, Tam CT. The use of quarry dust for SCC applications. CemConcr Res 2002;32:505–11.
15. Kwan AKH, Li Y. Effects of fly ash microsphere on rheology, adhesiveness and strength of mortar. Constr Build Mater 2013;42:137– 45.Methodology for the Mix Design of Self-Compacting Concrete... 123
16. Kwan AKH, Ng IYT. Improving performance and robustness of SCC by adding supplementary cementitious materials. Constr Build Mater 2010;24:2260–6.
17. Kwan AKH, Ng IYT. Optimum superplasticiser dosage and aggregate proportions for SCC. Mag Concr Res 2009;61(4):281–92.
18. Li LG, Kwan AKH. Mortar design based on water film thickness. Constr Build Mater 2011;25:2381–90.

REFERENCES

19. Nawa T, Izumi T, Edamatsu Y. State-of-the-art report on materials and design of self-compacting concrete. In: International workshop on self-compacting concrete; August 1998. p. 160–90.
20. Nepomuceno MCS, Pereira-de-Oliveira LA, Lopes SMR. Methodology for mix design of the mortar phase of SCC using different mineral additions in binary blends of powders. Constr Build Mater 2012;26:317–26.
21. Nepomuceno MCS, Pereira-de-Oliveira LA. Parameters for self-compacting concrete mortar phase. In: Fifth ACI/CANMET international conference on high-performance concrete structures and materials. Brazil, June 2008, ACI-SP-253-21, USA, May 2008, p. 323–40, ISBN: 978-0-87031-277-9.
22. Nepomuceno MCS. Methodology for self-compacting concrete mix design (in Portuguese). PhD thesis in civil engineering, University of Beira Interior, Portugal; 2005. p. 1–799. M.C.S. Nepomuceno et al. / Construction and Building Materials 64 (2014) 82–94 93
23. NP EN 12350-10:2010. Ensaios do betão no estado fresco. Parte 10: Betãoautocompactável. Ensaio de escoamentonacaixa L (in Portuguese). IPQ, Lisbon; 2010.
24. NP EN 12350-8:2010. Ensaios do betão no estado fresco. Parte 8: Betãoautocompactável. Ensaio de espalhamento (in Portuguese). IPQ, Lisbon; 2010.
25. NP EN 12350-9:2010. Ensaios do betão no estado fresco. Parte 9: Betãoautocompactável. Ensaio de escoamento no funil V (in Portuguese). IPQ, Lisbon; 2010.
26. NP EN 197-1:2001. Cement, Part 1: Composition, specifications and conformity criteria for common cements (in Portuguese). IPQ, Lisbon; 2001.
27. Okamura H, Ozawa K, Ouchi M. Self-compacting concrete. StructConcr J FIB 2000;1(1):3–17.
28. Ouchi M, Hibino M, Ozawa K, Okamura H. A rational mix-design method for mortar in self-compacting concrete. In: Proceedings of the sixth East-AsiaPacific conference on structural engineering & construction. Taiwan; 1998. p. 1307–12.
29. Pelova G, Takada K, Walraven J. Aspects of the development of self-compacting concrete in the Netherlands, applying the Japanese mix design system. In: Andreikiv OY, Luchko JJ, editors. Fracture mechanics and physics of construction materials and structures. 3rd ed., Kamaniar: The National Academy of Sciences of Ukraine; 1998.
30. Pereira-de-Oliveira LA, Nepomuceno MCS, Castro-Gomes JP, Vila MFC. Permeability properties of self-compacting concrete with coarse recycled aggregates. Constr Build Mater 2014;51:113–20.

31. Pereira-de-Oliveira LA, Nepomuceno MCS, Rangel M. An eco-friendly selfcompacting concrete with recycled coarse aggregates. Informes de la Construccion, vol. 65, n Extra-1; 2013. p. 31–41.
32. Petersson O, Billberg P, Bui Van. A model for self-compacting concrete. In: Proceedings of RILEM international conference on production methods and workability of fresh concrete. Paisley, London; June 1996. p. 484–92.
33. Petersson O, Billberg P. Investigation on blocking of self-compacting concrete with different maximum aggregate size and use of viscosity agent instead of filler. In: Proceedings of the 1st international RILEM symposium on SCC. Sweden; September 1999. p. 333–44.
34. Petersson O. Test method description: L-shape box test. In: Self-compacting concrete: state-of-the-art report of RILEM Technical Committee 174-SCC, RILEM; 2000. p. 126–8.
35. Sedran T, Larrard F. Optimization of self-compacting concrete thanks to packing model. In: 1st International RILEM symposium on SCC. Sweden; September 1999. p. 321–32.
36. Silva PMS, Brito J, Costa JM. Viability of two new mixture design methodologies for self-consolidating concrete. ACI Materials Journal, Title 108-M61, November 2011. p. 579–88.
37. Silva PMS, Brito J. Electrical resistivity and capillarity of self-compacting concrete with incorporation of fly ash and limestone filler. AdvConcrConstr 2013;1(1):65–84.
38. Silva PMS. Evaluation of durability of self-compacting concrete (in Portuguese). PhD thesis in civil engineering. Portugal: Technical University of Lisbon; 2013.
39. Sonebi M. Medium strength self-compacting concrete containing fly ash: modelling using factorial experimental plans. CemConcr Res 2004;34:1199–208.
40. Su N, Hsu KC, Chai HW. A simple mix design method for self-compacting concrete. CemConcr Res 2001;31(12):1799–808.Methodology for the Mix Design of Self-Compacting Concrete... 125
41. Su N, Miao B. A new method for the mix design of medium strength flowing concrete with low cement content. CemConcr Compos 2003;25(2):215–22.
42. Takada K, Tangtermsirikul S. Testing of fresh concrete. In: Self-compacting concrete: state-of-the-art report of RILEM Technical Committee 174-SCC. RILEM; 2000. p. 25–39.
43. Takada K. Test method description: box-shape test. In: Self-compacting concrete: state-of-the-art report of RILEM Technical Committee 174- SCC. RILEM; 2000. p. 123–25.

44. Takada K. Test method description: slump-flow test. In: Self-compacting concrete: state-of-the-art report of RILEM Technical Committee 174-SCC, RILEM; 2000. p. 117–119.
45. Takada K. Test method description: V-funnel test. In: Self-compacting concrete: state-of-the-art report of RILEM Technical Committee 174- SCC, RILEM; 2000. p. 120–122.
46. Tangtermsirikul S, Bui Van. Blocking criteria for aggregate phase of selfcompacting high-performance concrete. In: Proceedings of regional symposium on infrastructure development in civil Thailand engineering. Bangkok; December 1995. p. 58–69.
47. Topçu IB, Bilir T, Uygunoglu T. Effect of waste marble dust content as filler on properties of self-compacting concrete. Constr Build Mater 2009;23:1947–53.
48. Vieira MG. Self-compacting concrete – Rheology of concrete in the fresh state (in Portuguese). PhD thesis in civil engineering. Portugal: Technical University of Lisbon; 2008.

CITATION

Miguel C.S. Nepomuceno, L.A. Pereira-de-Oliveira, S.M.R. Lopes, Methodology for the mix design of self-compacting concrete using different mineral additions in binary blends of powders, Construction and Building Materials, Volume 64, 14 August 2014, Pages 82-94, ISSN 0950-0618, http://dx.doi.org/10.1016/j.conbuildmat.2014.04.021.

Chapter 3

Reliability Design Methodology for Confined High Pressure Inflatable Structures

E.J. Barberoa, E.M. Sosa[2], X. Martineza, J.M. Gutierrez[1]

[1]Mechanical and Aerospace Engineering, West Virginia University, Morgantown, WV 26506-6106, USA
[2]Civil and Environmental Engineering, West Virginia University, Morgantown, WV 26506-6106, USA

ABSTRACT

A design methodology for high pressure, inflatable structures is proposed. The inflatable structure may be partially confined inside large diameter conduits and tunnels. The design addresses the main structural requirements of the system, namely, fabric strength, geometric stability, and axial stability. The proposed design methodology is based on the concept of limit states. Load and resistance factors are identified for all the stochastic variables participating in the structural design equations. A formal methodology is used to estimate the load and resistance factors from known distributions of data for each of the stochastic variables. The concepts of basis values, coverage, and confidence are used along with the analytical treatment necessary to estimate the load and resistance factors. The analysis is applied to the cases of Normal, Log-normal, and

Weibull distributions of data. Material characterization and data analysis are presented for fabric strength and friction coefficient between the inflatable and the confining conduit material. The system reliability is also evaluated.

INTRODUCTION

Occasionally, large diameter conduits and tunnels need to be sealed to prevent the flow of liquids or gases in case of emergency, maintenance, or environmental remediation. Applications include large water and sewage pipes, large conduits for industrial fluids, rail and automobile tunnels, and so on that might be vulnerable to unexpected system failures that require temporary closure. For example, flooding of a freight tunnel in downtown Chicago and the buildings connected to these tunnels, forced evacuation of more than 250,000 people [1]. The Chicago tunnels have a cross section of 2.8 × 1.3 m and at the time of the accident they represented a network of more than 80 km that was used to run freight, television, telephone, and power conduits. Pumping water from the tunnel system took five and a half weeks at a cost of $40 million [2]. Although it is difficult to prevent all situations that can lead to such threatening events, damage can be substantially minimized by compartmentalizing the region affected by the event.

Inflatable structures, such as the one shown in Fig. 1, have been proposed [3], [4] and [5] for sealing large diameter conduits when it becomes necessary to block the flow of a pressurized liquid through the conduit as shown schematically in Fig. 2. Typically, the inflatable would be placed inside the conduit compactly folded inside a container (Fig. 1a) that holds the inflatable until a signal triggers the deployment (Fig. 1b); then, once the plug has deployed, the inflation process begins until the inflatable reaches its full shape (Fig. 1c and d); when the plug has reached the final shape, the pressurization starts and the inflatable is able to withstand the external pressure exerted by pressurized liquid acting on one side of the inflatable as illustrated in Fig. 2. The geometry is such that the inflatable fits 129 Reliability Design Methodology for Confined High Pressure.precisely the internal geometry of the conduit to be sealed, which is not necessarily that of a perfect cylinder. In addition to fitting the conduit, the geometry of the in inflatable includes the geometry of the end caps, which are to be designed to minimize the stress on the fabric.

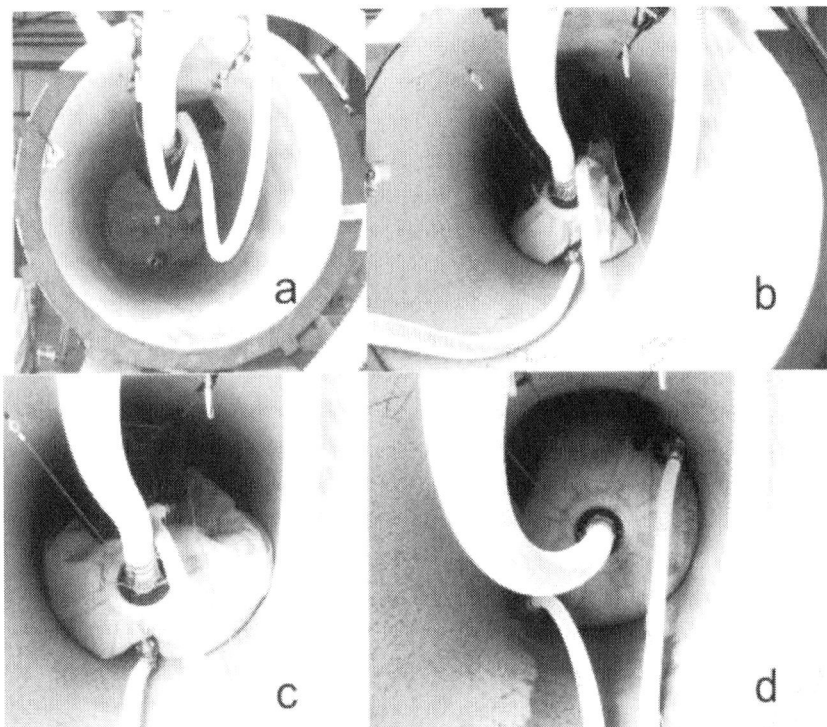

Figure 1: Inflation sequence (left to right). The fabric plug, initially hanging from the roof of the conduit (a), is filled with water through the large diameter hose (b). Any trapped air is released through the small diameter hose on top of the plug (c), which is then closed to allow for full pressurization (d). The small hose on the bottom of the plug is later used for draining the plug.

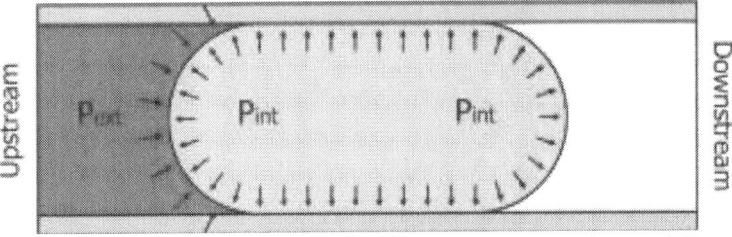

Figure 2: schematic of the proposed inflatable structure to seal a conduit.

The selection of the fabric hinges on the strength of the fabric and the strength of the welded, bonded, or heat-sealed fabric. Welds are necessary because the geometry of the inflatable is obtained by joining pieces of fabric cut appropriately to yield the desired geometry. Anchorage can be achieved by mechanical anchors or by friction between the inflatable and the conduit. The former introduces undesirably high tear stresses on the fabric at the anchorage points, requires intrusive preparation of the conduit via installation of anchors, and introduces additional modes of failures, including failure of the anchors and the anchoring points in the conduit. When a friction anchoring is used, the friction coefficient between the fabric and the conduit becomes a system property that needs to be characterized. In summary, the lowest fabric strength, welded or un-welded, plus the friction coefficient are the two *resistances* needed for the design.

Regarding the externally imposed loads, the pressure of the fluid to be contained p_e is specified by the particular application at hand. The inflation pressure p_i is a design parameter to be chosen by the designer to seek safe operation of the system. For the design presented here, the pressure p_e is assumed to be applied quasi-statically. No dynamic effects attributed to a sudden rush of fluid are explicitly considered in the design since the normal operation of the inflatable structure is expected to be carried out under quasi-static loading conditions. The importance factor used in the proposed design implicitly accounts for unexpected or unaccounted loading circumstances such as rushing fluid, among others. We acknowledge that the dynamic effects of a rushing fluid may affect the behavior of the structure at the early stages of operation, but that situation ruled out for the range of operating conditions considered in the present work.

While most inflatable structures in use today are inflated at low pressures (e.g., below 0.07 bars) the high pressure inflatables considered in this work are required to operate at substantially higher pressure, which is dependent on the external pressure that they must resist. For this reason, the inflatables discussed in this work are assumed to be constructed of a textile fabric capable of withstanding the stress imposed by the high inflation and external pressures. Such fabric is typically coated with a polymer to provide hermeticity but non-structural bladders may also be used for that purpose.Reliability Design Methodology for Confined High Pressure... 131

Limit States Design

Common practice for the design of inflatable structures uses allowable stress design (ASD) with large factors of safety to account for uncertainties. On the other hand, general procedures for limit states design have been developed for steel, concrete, wood, and masonry structures, for which significant data on performance under various loads, and load combinations exists, which have allowed resistance factors to be clearly defined. This is not the case for inflatable structures. The approach proposed herein is an attempt to improve the current methodology used by softgoods design engineers, whom for the most part are not familiar with reliability design.

The limit states design (LSD) method requires only one value to describe each stochastic variable (load, resistances, geometry, and so on). This single value is the *basis* value $x_{p,q}$, which for resistances, defines the interval ($x_{p,q}$, ∞) that contains a fixed proportion $f < 1$ of the population. In other words, it can be said with confidence q that $100f\%$ of the strength data will be above the basis value and that only $100p\%$ of the data will be below it, with $p = 1 - f$.

Since tolerance intervals are based on a sample **x** containing only n data points out of the entire population, the former assessment can be made only with a certain level of confidence $q < 1$. In the aerospace industry, A-basis is defined with $f = 0.99$, $q = 0.95$ and B-basis is defined with $f = 0.90$, $q = 0.95$. For some applications, such as wind turbine blades and marine structures, it is common practice to use a single tolerance interval, namely 95% coverage and 95% confidence [9], [10] and [11]. In this work, the basis values obtained with 95% coverage and 95% confidence are called C-basis.

Unlike allowable stress design (ASD), LSD uses multiple (partial) safety factors to take into account the uncertainty of each of the parameters separately (loads and resistances) [12] and [13]. This is in contrast to ASD where a single safety factor is used to safeguard against all the uncertainties without recognizing that various parameters may have drastically different uncertainties.

Although the basis values $x_{p,q}$ for each variable (loads and resistances) are the only values needed for the analysis, it is customary in LSD to use mean values and partial factors φ; one pair for each load and resistance. This is equivalent to splitting the basis value into a mean value of resistance R and a resistance factor φR using the definition

$$X_{p,q} = \varphi_R \text{ with } \varphi_R < 1 \tag{1}$$

For the loads, the upper tail of the distribution is used. In other words, the interval $(x_{p,q}, \infty)$ contains a fixed proportion $p = 1 - f$ of the population. Then, it can be said with confidence q that $100f\%$ of the load data will be below the basis value and that only $100p\%$ of the data will be above it. Again, the basis value is split into the the mean value of load L and a load factor αL using the definit

$$X_{p,q} = a_L L \text{ with } a_L > 1 \tag{2}$$

Furthermore, LSD recognizes two limit states: serviceability and ultimate. The inflatable structure is designed for the ultimate limit state [12] and [13]. That is, the structure is designed to safely sustain the loads that might cause failure of the inflatable system. The inflatable structure is not designed for serviceability criteria because the inflatable is not a structure meant to work in service (i.e., permanent condition), but rather it is a safety device that will deploy only during extraordinary events that are triggered by an emergency.

According to LSD, there are two complementary aspects to be considered separately: the resistances of the structure and the loads applied to it. Since there is no available LSD procedure for inflatable structures, as a starting point for this work, it is proposed that design must satisfy the following inequality [12]

$$\varphi_R R > {}_{\alpha D} D + \psi \gamma \{{}_{\alpha L} L + {}_{\alpha T} T\} \tag{3}$$

where ϕR is the resistance factor, R the material resistance, D the dead load, L the live load, T the thermal load, αD the dead load factor, αL the live load factor, αT the thermal effect (temperature) load factor, ψ the load combination factor and γ is the importance factor.

As shown in Fig. 2, there are two main loads acting over the inflatable structure, the inflation pressure and the external pressure. Both of these are live1 loads. There are no dead loads.2 No thermal loads have been defined for this application, but they may be included in the future if the fluids being transported by the conduit are not at room temperature.

The two loads applied to the structure define three different limit states that must be verified using (3). These ensure that the structural strength of the material is sufficient to resist the stresses, that the axial stability of the structure is maintained, and an additional functional requirement that relates the internal and external pressure applied to the inflatable plug.

For this particular application, the importance factor is determined by the criticality of failure of the system and the existence or not of redundant loads paths to alleviate unexpected circumstances. In addition, for this particular application, *customers specifications* require us to include an factor γ to account for the consequences of the system failing due to unexpected or unaccounted circumstances. The influence of the importance factor value chosen on the system reliability in evaluated in Section 5.

Determination of the Inflation Pressure

The inflation pressure pi must be larger than the fluid pressure pe to maintain the upstream end cap geometrically stable (Fig. 2). Furthermore, if pi drops closer to pe, the fluid can seep around the inflatable creating a layer that reduces the contact between the inflatable and the conduit walls drastically, thus compromising the anchoring force, which is entirely due to friction. On the other hand, the inflation pressure should be as small as possible to minimize the stresses in the fabric on the downstream cap (Fig. 2). This latter requirement is now formalized as follows:

$$\phi_{pi} p_i > \psi \gamma_{\alpha L} p_e \tag{4}$$

Note that the inflation pressure is on the *resistance* side of the inequality (4) and that a reduction of that pressure would be detrimental. Therefore, the inflation pressure is multiplied by a *reduction* factor $\phi_{pi} \leq 1$. This is in contrast to the fluid pressure *pe*, which is on the *load* side of the inequality (4) and thus it is affected by a load factor $\alpha_L \geq 1$.

The *resistance factor* ϕ_{pi} is determined by the variability in the inflation pressure *pi*, which is maintained by a pump, valve, and control system with variability described by a Normal distribution. Therefore, the basis value for the resistance factor ϕ_{pi} is calculated according to the methodology presented in [14] as:

$$X_{p,q} = \bar{x} - k_{p,q}(n)_s \tag{5}$$

where, $\bar{x}, s$ are the mean value and standard deviation of the sample data, and $k_{p,q}$ is tabulated in [14, Table 1] as a function of the selected coverage *f*, confidence *q*, and sample size *n*. Usually, a sufficiently large data set ($n \to \infty$) exists for the pressure provided and controlled by inflation system, which allows one to assume that the sample mean $\bar{x}$ and standard deviation *s* approach the population mean and variance σ. In other words, the population mean and variance are known with high confidence (i.e., *q*= 1), so that $\bar{x} \to \bar{\bar{x}}$ and $s \to \sigma$. In this case, discarding the lower tail of the distribution:

$$\bar{k_{p'1}} = \Phi^{-1}(f) \tag{6}$$

that is, the factor can be calculated simply as the value of the inverse cumulative distribution function in standard form [15] and [16], which is widely available in tabular form, or using the MATLAB® function *icdf*('normal', *f*, 0, 1), where *f* is the coverage (e.g., *f* = 0.95 for C-basis). Combining (2), (5) and (6), the resistance factor can be calculated as:

$$\varphi_{pi} = 1 - \Phi^{-1}(f)_{CV} \tag{7}$$

where $C_v = s/\bar{x}$ is the coefficient of variation (COV).

The *load factor* αL to be applied on *pe* is chosen taking into account that the upstream pressure is a live load[17]. The upstream pressure is due to the fluid head of a reservoir with variability typically described by a Normal distribution. Then, the basis value is calculated according to (5). Usually, a sufficiently large data set ($n \to \infty$) exists for the reservoir head (i.e., pressure head due to the level of the reservoir as a function of time, from historical records). That allows one to assume that the population mean and variance are known with high confidence (i.e., $q = 1$). In this case, discarding the higher tail of the distribution

$$k_{p'1} = \Phi^{-1}(p) \tag{8}$$

that is, the factor can be calculated simply as the value of the inverse cumulative distribution function in standard form [15] and [16], which is widely available in tabular form, or using the MATLAB® function *icdf*('normal', *p*, 0,1), where $p = 1 - f$, and f is the desired coverage (e.g., $f = 0.95$, $p = 0.05$, for C-basis). Combining ((2), (5) and (8)) and noting that $\Phi^{-1}(p) = -\Phi^{-1}(f)$, the load factor can be calculated for Normally distributed data as

$$\alpha L = 1 + \Phi^{-1}(f)_{CV} \tag{9}$$

Since there is no dead load, the load combination factor is $\psi = 1.0$.

The importance factor γ is chosen taking into account that violating (4) may result in partial loss of frictional anchoring force, which may partially imperil the functionality of the system. The selection of the importance factor for the inflation pressure is based on the ability of the system to recover from unexpected loss of pressurization, which is typically achieved having redundant detection and inflation systems that can back up the primary systems in case of malfunction and that are activated before complete depressurization of the inflatable. The importance factor γ also takes into account whether on not redundant features and/or redundant load paths exist that can mitigate the effect of unexpected overloads. Fabric failure may be caused by an unexpected rise of fabric stresses, for example due to unaccounted stress concentrations attributed to plug misalignment, plug distortions, lack of conformance to the conduit and so on, which if they exceed the strength of the material would lead to a catastrophic failure of the whole system. Also, unexpected changes in the surface roughness and/or cleanliness of the conduit's surface leading to changes in the friction coefficient, would compromise the axial stability, meaning that the inflatable would slide along the conduit, with the consequent catastrophic failure of the system. Finally, the required inflation pressure is calculated from (4) as

$$p_i > \frac{\gamma a_L P_e}{\varphi p_i} \tag{10}$$

Design for Material Strength

An inflation pressure *pi* is applied to inflate the structure sufficiently for it to maintain its shape and to generate enough friction against the conduit surface to be able to prevent the inflatable from sliding along the conduit under the action of the upstream fluid pressure *pe*.

On the portion of the inflatable that is in contact with the wall (Fig. 3), heretofore called the cylindrical portion, the wall equilibrates the inflation pressure exactly and the fabric is subject to negligible stress. On the cylindrical portion, the fabric acts only as an impervious media and helps contain the inflation fluid during the inflation process, but once inflated, the fabric in the cylindrical portion is subjected to no appreciable stress. On the upstream side, the fabric is subjected to the stress necessary to equilibrate the pressure differential between the upstream pressure *pe* and the inflation pressure *pi*. On the downstream side, the fabric holds the stress needed to equilibrate the entirety of the inflation pressure *pi*. For this reason, *pi* is chosen as small as possible.

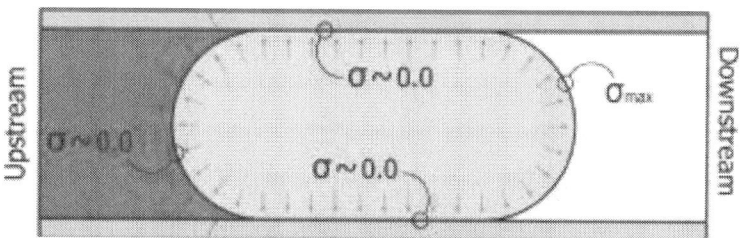

Figure 3: Stresses in the different regions of the inflatable.

For calculation of stress in the fabric, the fabric can be assumed to be inextensible. This is a good approximation for the type of fabrics used and the pressures that can be applied to them. Therefore, the geometry does not change appreciably as a result of inflation and the calculation of stress in the fabric is that of an isostatic structure. The geometry of the membrane has, at every point, two principal radii of curvature $r1$ and $r2$. According to Laplace's equation [18], the largest principal stress-resultant is

$$N = t\sigma = \frac{\Delta P}{(1/r_1 + 1/r_2)} \tag{11}$$

Where t is the thickness of the fabric and ΔP is the pressure differential across the membrane (i.e., the pressure differential contained by the membrane). For fabrics, it is customary to use stress-resultants $N = t\sigma$ in N/m instead of stress σ. For a sphere, $r1 = r2 = r$, and for a cylinder, $r1 = r$, $r2 = \infty$, so

$$N^{sphere} = \frac{r}{2}\Delta P \tag{12}$$

$$Ncyl = r\Delta P \tag{13}$$

Comparing (12) with (13), one concludes that the stress-resultant in a membrane of general curvature is bracketed by the values for the sphere and cylinder, with the value for the cylinder being the maximum possible. In our current application, Fig. 3 shows that the loads in the cylinder are directly transferred to the conduit surface and, therefore, the maximum fabric stress-resultant is found at the downstream sphere, which is defined by Eq. (12), with $\Delta P = pi$, where pi is defined by (10). According to the limit state Eq. (3),

$$\varphi_R R > \psi_{\alpha L} N \qquad (14)$$

Where R is the mean value of resistance of the fabric and φR is the resistance factor associated to the value of R. This factor depends on the variability of R. The fabric strength data is analyzed in Section 3.

The stress-resultant N is calculated using (12). The load factor αL to be applied on pi in (14) is chosen taking into account that the inflation pressure is a live load with variability given by the inflation system (pumps, regulators).

Since there is no dead load, the load combination factor is chosen to be $\psi = 1.0$. From (12), (13) and (14) one can calculate the required fabric resistance as

$$R > \frac{a_L r p_i}{2\varphi_R} \qquad (15)$$

Design for Axial Stability

The conduit is anticipated to a have a prismatic cross section, not necessarily circular, with perimeter S. The inflatable geometry mimics the prismatic geometry of the conduit and it is in contact with it over a contact length Lc, defined by the length of the prismatic section. Therefore, the axial force that friction can sustain can be calculated as follows:

$$Fr = \mu_{piSLc} \qquad (16)$$

Where μ, pi, S, Lc, are the friction coefficient, inflation pressure, perimeter, and contact length, respectively. For a circular conduit of radius r, the perimeter is $S = 2\pi r$.

The applied axial force is caused by the pressure differential between the upstream pressure p_e and downstream pressure p_o. Assuming atmospheric downstream pressure, the applied axial load is

$$F_a = p_e A \qquad (17)$$

Where A is the cross-sectional area of the conduit. Both A and S are assumed to be deterministic parameters (i.e., they have no variability). For a circular conduit of radius r, the area is $A = \pi r^2$.

According to the LSD Eq. (3),

$$\varphi F F_r > \psi_\alpha L F_a \qquad (18)$$

From (16), (17) and (18), the following limit state requirement is obtained

$$\varphi_\mu \mu \varphi_{pi} p_i S L > \psi_{\alpha pe} p_e A \qquad (19)$$

where p_i is defined by (10) and the resistance factor φ_μ is determined by the variability of the friction coefficient μ, analyzed in Section 3.

Since there is no dead load, the combination factor is chosen again as $\psi = 1$.

From (19) one can calculate the contact length required for the inflatable as

$$L_c > \left(\frac{a_{pe} P_e}{\varphi_{pi} P_i} \right) \left(\frac{r}{2 \varphi_\mu \mu} \right) \qquad (20)$$

MATERIAL CHARACTERIZATION

According to LSD [17], there are two complementary aspects to be considered: the resistances of the structure and the loads applied to it. There are two material/system properties, or *resistances*, that control the resistance aspect of the design: the tensile strength of the fabric R and the friction coefficient μ between fabric and conduit.

Tensile Strength of the Fabric

A variety of fabrics can be used for inflatable structures as long as they are coated by a polymer. The coating serves two purposes. First, it makes the fabric impervious to contain the inflation pressure. Second, it provides a weldable substance, so that cut pieces of fabric can be welded together into the desired shape. Other joining techniques, such as stitching, do not provide as high load transfer as welding, and they are slower and more expensive to execute. Un-coated fabrics can still be welded by local insertion of a weldable polymer that impregnates and welds the fabric but the un-coated fabric would lack imperviousness. Fabrics can be coated on one or both sides. When coated on one side only, a strip of coating needs to be used to provide enough weldable polymer to impregnate and weld effectively the un-coated side of the fabric.

Four types of fabrics were considered for manufacturing the prototype designed in this work. The main properties of the each fabric considered are summarized in Table 1. The most important requirements for the design are that the prototype is required to sustain the membrane stresses exerted by the inflation and fluid pressure, as well as to provide acceptable friction properties to assure axial stability of the structure. After preliminary evaluation of the options shown in Table 1, the fabric chosen for the manufacturing of the prototype designed in this work is a woven fabric, constructed of Vectran HS fibers, 1500 denier with 4 × 4 basket weave pattern. This fabric is characterized by a total of 34 yarns per 25.4 mm the warp direction and 42 yarns per 25.4 mm in the fill direction. The thickness is 0.838 mm. This fabric had also a urethane polymer coating on both sides [19] and [20].

Table 1: Fabric materials considered for the design

Manufacturer	Identification	Base	Coating	Weave	Tens. strength warp/fill (N/mm)
Ferrari	Precontraint	Woven polyester	PVDF coating	Plain weave	77/76
	1002 Formula S		Both sides		
Seaman Corp.	Style 7150 PFF	Woven Nylon	Polyether	Plain weave	131/131
			Polyurethane		
			Both sides		
Warwick Mills	4 × 4 Vectran	Woven Vectran	Urethane polymer	Basket weave	406/450
			Both sides		
Warwick Mills	2 × 2 Vectran	Woven Vectran	Urethane polymer	Plain weave	400/400
			One side		

Tensile strength of fabrics R is reported as a force per unit width (i.e., [N/mm]). Tensile tests were performed on samples of fabrics to generate enough data to support the development of the resistance factor used in the design. Previous research [21] have demonstrated that the biaxial strength of coated fabrics is equal to the uniaxial strength. Therefore, the tensile strength of the material used in this work was characterized by standard uniaxial tests. Specifically, tensile tests were performed according to ASTM D 5035 [22], on a universal testing machine, model MTS 318.10 with hydraulic wedge grips specialized to hold fabrics, model MTS 647.10A [20]. The data is analyzed in Table 2 for the purpose of this study.

MATERIAL CHARACTERIZATION

Table 2: Tensile strength of urethane impregnated/coated, 4 × 4 basket weave, Vectran 1500 denier

		Specimen	Fill direction R (N/mm)	Warp direction R (N/mm)
		1	443.246	387.030
		2	444.122	418.203
		3	468.815	397.188
		4	444.647	437.467
		5	447.799	363.038
		6	480.723	418.203
		7	441.495	437.292
		8	444.122	397.188
		9	412.599	405.419
		10	467.939	401.040
Mean (N/mm)	$\overline{R}$		449.551	406.207
mle shape	$\hat{\lambda}$		458.141	416.264
mle scale	$\hat{k}$		27.080	21.441
Data points	n		10	10
Coverage	f		0.95	0.95
Confidence	q		0.95	0.95
From table [14]	$Vpqn$		8.751	8.751
Basis value	xpq		370.668	318.532
Resist factor	φR		0.825	0.784

Fabric strength is highly dependent on fiber strength, which is known to be best represented by a Weibull distribution. Therefore, the basis value is calculated using [14]

$$x_{p,q} = \hat{\lambda}[-In(1-p)]1/\hat{k} \exp\left(\frac{-V_{p,q,n}}{\hat{k}\sqrt{n}}\right) \qquad (21)$$

Where $\hat{\lambda}, \hat{k}$ are the maximum likelihood estimates (*mle*) for the data and $V_{p,q,n}$ is tabulated in [14] as a function of the coverage *f*, confidence *q*, and sample size *n*. The calculations are summarized in Table 2, with the *mle* values calculated using the MATLAB function *mle(x, 'distribution', 'Weibull')*, and *x* is the data.

From Table 2, the lowest basis value is $x_{95,95}$ = 318.532 N/mm for the warp direction of the fabric, corresponding to a mean value $\overline{R}$ =406.207 N/mm and resistance factor ɸ*R*= 0.784

Friction Coefficient

The friction coefficient μ is a system property that controls the amount of axial force that can be obtained from a given inflation pressure. It is assumed that the inflatable is in contact with the internal surface of the conduit, which could be concrete, masonry, steel, or other. The surface could be smooth or rough depending on the location where the inflatable is installed. Also, the fabric and the surface of the conduit may be dry or wet when the inflatable is deployed. Tests were performed to investigate these configurations.

The friction coefficient is a property that is strongly dependent on the system configuration [23]; that is, type of surfaces, combination of loads applied to the surfaces, lubrication effect of liquids if present, and so on. Therefore, a custom test set up was developed for determination of the friction coefficient [19]. The experimental apparatus is shown in Fig. 4. A hydraulic pump (1) pushes a car with a concrete block on it (2). The car slides below a piece of fabric (3) that is held by the vertical loading ram, loaded with a known weight (4). The square fabric specimen has a surface area of 12.65 cm2 and it is held to the flat end of the loading ram by double-side adhesive tape. The horizontal force required to move the car at constant speed is measured by a load cell (5). The displacement of the car is measured with a linear voltage differential transducer (LVDT) (6). The load cell and LVDT output are converted to SI units and recorded on a computer via a data acquisition system.

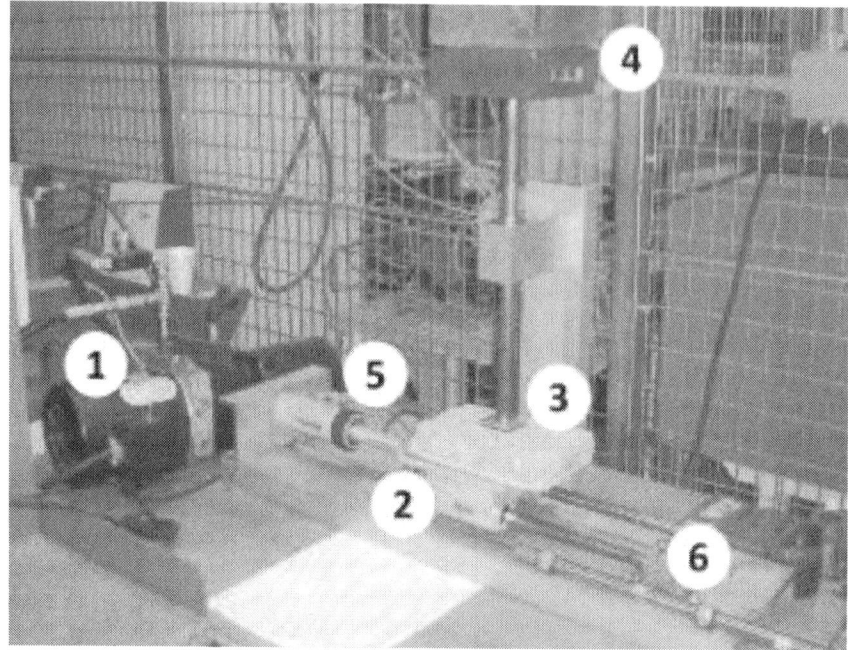

Figure 4: Apparatus used to measure the friction coefficient. 1: Pump, 2: carriage, 3: sample holder, 4: weight, 5: load cell, 6: displacement transduced.

To reproduce with the maximum possible accuracy the configuration of our system, the vertical load applied to measure the friction coefficient is 309.0 N, which gives a normal pressure of 239 kPa. Friction coefficient data for friction between Urethane-coated 4 × 4 Vectran and rough and smooth concrete under dry and wet conditions from [20] is analyzed in Table 3 for the purpose of this work.

Table 3: Friction coefficient between concrete and urethane coated Vectran (1500 denier, 4 × 4 basket weave), applied pressure 239 kPa, and velocity 4.2 cm/s.

Specimen		Concrete			
		Smooth		Rough	
		Dry	Wet	Dry	Wet
		μ	μ	μ	μ
1		0.668	0.541	0.616	0.55
2		0.67	0.546	0.668	0.54
3		0.74	0.628	0.619	0.522
4		0.721	0.692	0.641	0.552
5		0.751	0.679	0.595	0.533
Mean	$\bar{\mu}$	0.71	0.617	0.628	0.539
Mean ln	xLN	−0.343	−0.488	−0.466	−0.617
Standard dev. ln	sLN	0.055	0.117	0.044	0.023
Data points	n	5	5	5	5
Coverage	f	0.95	0.95	0.95	0.95
Confidence	q	0.95	0.95	0.95	0.95
From table [14]	$kp\,q$	4.203	4.203	4.203	4.203
Basis value	$xp\,q$	0.563	0.375	0.522	0.49
Resist factor	$\varphi\mu$	0.793	0.608	0.831	0.909

Since friction cannot be negative, the friction data is assumed to have a Log-normal distribution. Therefore, the basis value is calculated using [14], as:

$$x_{p,q} = \exp[\bar{x}_{LN} k_{p,q}(n) s_{LN}] \tag{22}$$

Where $\bar{x}_{LN}$ is the sample mean in log space, $3sLN$ is the standard deviation in log space, and $kp,q(n)$ is tabulated in [14] as a function of the coverage f, confidence q, and sample size n. The calculations are summarized in Table 3.

PROTOTYPE DESIGN

In this section, the methodology proposed in Sections 2 and 3 is applied to the design of a prototype. The specifications for the prototype are:

- Circular cylindrical shape.
- Test pipe radius rp= 0.61 m.
- Upstream pressure pe= 207 kPa.

Although the test pipe is circular with radius r = 0.61 m, the radius of the inflatable was chosen slightly larger (ri= 0.635 m) to Reliability Design Methodology for Confined High Pressure... 145
ensure that the inflatable fits snugly against the pipe even if some wrinkles develop during inflation.

Inflation Pressure

The inflation pressure is determined using (10) for a given upstream pressure pe. The upstream pressure is assumed to have a Normal distribution with COV = 5%. Using (8) yields a load factor αL = 1.082. The inflation pressure is assumed to have a Normal distribution with COV = 10%. Using (6) yields a resistance factor ϕpi=0.872. The importance factor is chosen to be γ = 1.2 to be sure that the nominal inflation pressure is at least 20% higher than the nominal fluid pressure. Then,

$$p_i > \frac{\gamma a_L p_e}{\varphi_{pi}} = \frac{1.2 \times 1.082 \times 207}{0.872} = 3.8.221 kPa \qquad (23)$$

Therefore, the inflation pressure is chosen to be pi = 310 kPa.

Fabric Strength

The required material strength R is determined using (15) for a given pipe radius r_p and selected inflation pressure p_i. Since p_i acts as a load, its load factor is calculated with (8), i.e., $α_L$ = 1.128.

The strength data collected in this study was analyzed in Section 3.1. Considering a C-basis, the resistance factor was calculated to be $φ_R$= 0.784.

Since only one load is considered, the load combination factor is ψ = 1. Then, from (15), the required fabric strength is

$$R > \frac{a_L r p_i}{\varphi_R} = \frac{1.128 \times 0.635 \times 310}{2 \times 0.784} = 141.611 N / mm \qquad (24)$$

From Table 2, the available fabric strength is R = 406.207 N/mm, which is sufficient for this application.

Length of the Inflatable

The length required to develop sufficient axial resistance to prevent the inflatable from sliding under the action of the upstream pressure is calculated with (20). Since the test pipe is cylindrical, the cross section is $A = πr^2$ = 1.169 m2 and the perimeter is $S = 2πr$ = 3.833 m.

The friction coefficient data collected in this study was analyzed in Section 3.2. The mean value of friction is μ= 0.617. Considering C-basis, the friction resistance factors is φμ= 0.608.

Since the inflation pressure acts as a resistance, its resistance factor is calculated with (6), i.e., φpi=0.872. On the other hand, the upstream pressure acts as a load, so its load factor is calculated with(8), i.e., αpe=1.082. Therefore,

$$R > \frac{a_L r p_i}{\varphi_R} = \frac{1.128 \times 0.635 \times 310}{2 \times 0.784} = 141.611 N / mm \qquad (25)$$

Due to manufacturing reasons, the length of the cylindrical portion is chosen to be L_c= 0.70 m.

SYSTEM RELIABILITY

The confined inflatable plug can be considered as a series system. In order to work, the three criteria (i.e. inflation pressure, axial stability, and material strength) have to be met simultaneously. The failure functions that describe this system can be written as

$$g1 = \overline{p_i} - \overline{p_e} \tag{26}$$

$$g2 = 2 L_{cpi} \mu - p_{er1} \tag{27}$$

$$g_3 = R - \frac{p_i r_2}{2} \tag{28}$$

where $g1$, $g2$, $g3$ are the failure functions for the inflation pressure, axial stability and fabric strength, respectively. The description of each one of the variables involved in the failure functions is given in Table 4. It is assumed that the statistical parameters calculated from Reliability Design Methodology for Confined High Pressure... 147 experimental data are the parameters of the probability distribution functions for friction and fabric strength.

Table 4: Variables definition for reliability analysis.

Variable	Description	Distribution	Parameter 1	Parameter 2
r_1	Test pipe radius	Deterministic	610 mm	-
r_2	Plug hemisphere radius	Deterministic	635 mm	-
Lc	Cyl. portion length	Deterministic	700 mm	-
pi	Inflation pressure	Normal	Mean = 0.310 MPa	COV = 0.10
pe	Upstream pressure	Normal	Mean = 0.207 MPa	COV = 0.05
R	Fabric strength (warp)	Weibull	λ = 416.264 N/mm	k = 21.441
μ	Fric. coeff. (smooth-wet)	Log-normal	Mean LN = -0.488	STDEV LN = 0.117

Inflation Pressure

Interference theory [26] and [27] is used to calculate the probability of failure for the failure function g1. The function includes a linear combination of two random variables with Normal distribution for which the probability of failure is given by

$$\overline{\mu}_{gl} = 308.22 - 207 = 101.22 kPa \tag{29}$$

$$\sigma_{gl} = \sqrt{\sigma_{pi}^2 = \sigma_{pe}^2}$$
$$\sigma_{gl} = \sqrt{30.822^2 + 10.35^2} = 32.51\, kPa \tag{30}$$

Where $\overline{\mu}_{gl}, \sigma_{gl}$ are the mean value and standard deviation of the failure function, respectively. Therefore, the probability of failure(i.e. P{g1 ⩽0}) for the inflation pressure requirement can be calculated as

$$P_{F_{gl}} = \phi\left(0, \overline{\mu}_{g3}, \sigma_{g3}\right)$$
$$P_{F_{gl}} = \phi\left(0, 101.22, 32.51\right) = 9.354 \times 10^{-4} \tag{31}$$

Then, the reliability for this condition of the system is Rg1=0.999065.

Axial Stability

The axial stability failure function is a non-linear combination of Normal and Log-normal random variables. For this analysis, three variables are considered as stochastic: the inflation pressure pi, the friction coefficient μ, and the upstream pressure pe. The distribution types and parameters used for the computations are listed in Table 4. The probability of failure is computed using the gradient based algorithm proposed by Rackwitz and Flessler [24]. The estimated probability of failure is

$$PF_{g2} = 2.181 \times 10^{-5} \quad (32)$$

which corresponds to a reliability of Rg2=0.999978.

Material Strength

The failure equation that represents material strength is a function of two random variables: the fabric strength R (Weibull distribution) and the inflation pressure pi (Normal distribution). Applying interference theory [26] and [27], the probability of failure is

$$PF_{g3} = 2.492 \times 10^{-13} \quad (33)$$

Which corresponds to a reliability of practically Rg3≈1.0.

Series System

For *series* systems made of n components, the probability of failure of the system PF, can be bounded using the individual probabilities of failure of each component Pi [28]

$$\max\{P_i\} \leq P_F \leq \sum_{i=1}^{n} P_i \quad (34)$$

where the lower limit is determined by the maximum of the individual probabilities of failure, and the upper limit is defined by the sum of the individual probabilities of failure. Then, substituting the individual probabilities of failure obtained in (31), (32) and (33) in (34), yields

$$9.354 \ 10^{-4} \leq P_F \leq 9.572 \ 10^{-4} \tag{35}$$

Therefore the reliability R of the system is bounded by

$$0.999043 \leq \Re \leq 0.999065 \tag{36}$$

Influence of Importance Factor on System Reliability

It can be seen in (26), (27) and (28), that there are two sets of variables that affect the design. The first set includes the external pressure *pe*, the fabric strength *R*, the friction coefficient µ, and the radius of the confining surface *r*. Typically, the characteristics of this first set of variables are predetermined and cannot be easily changed. The second set of variables includes the inflation pressure *pi*, and the contact length *Lc*. From this second set, the length *Lc* is function of the inflation pressure as seen in (25). Therefore, the main variable that the designer can adjust during the design process is the inflation pressure *pi*.

The inflation pressure *pi* is function of a given upstream pressure *pe* and of a selected importance factor γ, as seen in (23). Typical importance factors for structures subjected to wind pressures vary from 1.0 to 1.15 depending of the category of the building [25] or from 1 to 1.5 for seismic design depending of the occupancy category defined by the code [25]. Since there are no codes for the design of confined inflatable structures, it is important to evaluate the influence of the importance factor γ on the individual probabilities of failure described in Sections 5.2 and 5.3. Three values of importance factors were chosen to evaluate such influence: γ = 1.0, 1.2, 1.5, and the reliability (as well as probability of failure) were evaluated for each failure function and for the complete system.

For failure function g1, the probabilities of failure show that as the importance factor increases, the probability of failure decreases two orders of magnitude for each increase in the importance factor, as summarized in Table 5. Note that for γ = 1.0 the reliability is 96.41%, which given the criticality of the structure, may be deemed insufficient. However, when γ = 1.2 the reliability approaches to 99.91%, and for γ= 1.5 the reliability increases to 99.9996%. Since the later may be unfeasible to achieve from a technical point of view, an importance factor for the pressure of γ = 1.2 is considered appropriate for this design.

Table 5: Influence of importance factor on failure equation g1

$g_1 = pi - pe$	Importance factor for pi (Eq. (23))		
	1	1.2	1.5
Mean pe (kPa)	207.00	207.00	207.00
Mean pi (kPa)	256.85	308.22	385.28
Std. dev. pe (kPa)	10.35	10.35	10.35
Std. dev. pi (kPa)	25.69	30.82	38.53
Reliability $_{Rg_1}$	0.964070	0.999065	0.999996
Probability of failure PFg_1	0.035930	0.000935	0.000004

For failure function g2, the change of the length Lc calculated using (25), as function of different values of *pi*, does not seem to affect the reliability as the importance factor increases. For all the values of γ, the reliability values remained in the same order of magnitude, as summarized in Table 6.

Table 6: Influence of importance factor on failure equation $g2$

$g_2 = 2Lcpi\mu - per$	Importance factor for pi (Eq. (23))		
	1	1.2	1.5
Mean pi (kPa)	256.85	308.22	385.28
Std. dev. pi (kPa)	25.69	30.82	38.53
Length Lc (mm) (Eq. (25))	813	677	542
Reliability $_{Rg2}$	0.999978	0.999978	0.999978
Probability of failure PFg_2	0.000022	0.000022	0.000022

For failure function $g3$, the stress is calculated using (24) as function of different values of pi. As expected, the reliability decreases as the importance factor increases because the stress approaches the strength of the fabric, thus reducing the margin of safety. For each increase of the importance factor, the probability of failure increases two orders of magnitude. Despite this increase, the probabilities of failure are still very small and the resultant reliabilities are practically 100%. These results are illustrated in Table 7.

Table 7: Influence of importance factor on failure equation $g3$

$g_3 = R - pir/2$	Importance factor for pi (Eq. (23))		
	1	1.2	1.5
Mean pi (kPa)	256.85	308.22	385.28
Std. dev. pi (kPa)	25.69	30.82	38.53
Stress (kPa) (Eq. (24))	117.33	140.80	176.00
Reliability $_{Rg3}$	1.00	1.00	1.00
Probability of failure PFg_3	4.446×10^{-15}	2.492×10^{-13}	2.579×10^{-11}

The probability of failure of the system defined in (34) is used again to determine the limits as function of the importance factor. Table 8 summarizes these limits based on the results obtained from the individual failure equations presented in Table 5, Table 6 and Table 7. As seen in Table 8, the reliability of the system is bounded by a very narrow range for each individual importance factor and increases as the importance factor increases.

Table 8: Limits of probability of failure and reliability of the system as function of the importance factor

Limits for the whole system	Importance factor for pi		
	1	1.2	1.5
Probability of failure			
Lower limit	0.035930	0.000935	0.000022
Upper limit	0.035952	0.000957	0.000026
Reliability			
Upper limit	0.964070	0.999065	0.999978
Lower limit	0.964048	0.999043	0.999974

A more general approach can be used for the reliability design of confined inflatable structures as proposed by Galambos et al. [6] and Ellingwood et al. [7] based on the classical reliability theory presented by Ang and Allin [8]. This general procedure has been used to determine load combination factors for different reliability indexes that are currently used in design codes such as ACI or AISC. However, implementation of the general methodology for the design of inflatable structures is not a simple task. In the general approach, the random variables with non-Normal distributions require transformation from non-Normal variables to equivalent Normal variables prior to the solution of the limit state Eqs. (26), (27) and (28). Then, the reliability of design can be evaluated by [6], [7] and [8]

$$g(\phi_1\mu_1, \cdots, \phi_n\mu_n) = 0$$

$$\phi_i = 1 - \alpha_i \beta \frac{\sigma_i}{\bar{\mu}_i}$$

$$\alpha_i = \frac{\sigma_i \frac{\partial g}{\partial x_i}}{\sqrt{\sum \left(\sigma_i \frac{\partial g}{\partial x_i}\right)^2}}$$

(37)

Where $\bar{\mu}_i, \sigma_i,$ are the mean and standard deviation of µi, respectively; β is the target reliability, also called reliability index or safety index. This index, according to Galambos et al. [6] "... may be established by reviewing the levels of reliability inherent in those existing standards which have resulted in the past in satisfactory performance". While target reliabilities are available reinforced concrete, steel, wood, composites, and masonry, there are not available for *confined* inflatable structures and there are very few structures of this type to look at. Thus, there are no target reliabilities available for this new type of structures. The designer would have to assume target values that will need to be confirmed either by experiment or by numerical simulations in order to find optimum or minimum values of β for satisfactory designs.

Note also that the geometrical cap stability Eq. (26) and the material strength Eq. (28) are linear equations but the axial stability Eq. (27) is nonlinear. For the implementation of the general method, the later will require linearization of the failure surface. Otherwise, the selected reliability index β could be affected by the formulation of the limit state equation [6]. Since the factor α_i is a function of the gradient ∂g/∂xi and that(27) includes the product of two random variables (µ and *pi*), the gradient with respect to one variable depends on the other variable. Therefore, the solution of (37) is not trivial. In summary, the general methodology is more complex and requires information that is not available at this early stage of development of confined inflatable structures, thus making its implementation to exceed the scope of our effort.

CONCLUSIONS

A reliability based methodology is proposed for the design of *confined* inflatable structures. Determination of resistance factors has been shown to follow naturally from the variability of the experimental data normally gathered in support of this type of design. The limit states design procedure helps the designer pay due consideration to each and every one of the various important aspects that may compromise the safety of the system. While the selection of resistance factors falls mostly on the designer, the selection of load factors and importance factor also involves the customer, helping them focus on what is important. This is not always an easy task, but it is crucial for the success of the application. By analyzing the friction coefficient data, it is shown empirically that the basis value governs the design, rather than the mean or the resistance factor. This further demonstrates that the limit states design methodology in this work is a Level I reliability method, which uses only one value (the basis value) for each stochastic variable in the design equations. Further, it is shown that the basis values can be calculated easily and directly from the experimental data that normally accompanies the design effort for novel applications of inflatables at high pressure, confined situations. The reliability of the design is evaluated in terms of a single importance factor. Results show that although individual probabilities of failure are different for each failure condition and have different orders of magnitude, the system reliability increases as the importance factor increases.

ACKNOWLEDGEMENTS

This work was funded by the U.S. Department of Homeland Security Science and Technology Directorate. The support of the U.S. Department of Homeland Security and the U.S. Transportation Security Administration is gratefully acknowledged.

REFERENCES

1. Inouye RR, Jacobazzi JD. The great Chicago flood of 1992. ASCE 1992;62(11):52–5.
2. Federal Transit Administration. Transportation security, TCRP report 86 and NCHRP report 525, vol. 12; 2006. p. 37–39.
3. Stocking AW. An inflatable tunnel seal stops flooding of world's largest undeveloped uranium mine; 2009. <http://www.petersenproducts.com/case_study/Large_Mine_Flooding_Remediation.aspx>.
4. Martinez X, Davalos JF, Barbero EJ, Sosa EM, Huebsch WW, Means K, et al. Inflatable plug for threat mitigation in transportation tunnels. In: SAMPE 2012, Baltimore, MD; May 21–24, 2012.
5. Web resource. <http://www.mae.wvu/barbero/research/RTP.html>.
6. Galambos TV, Ellingwood B, MacGregor J, Cornell CA. Probability based load criteria: assessment of current design practice. J StructDiv, ASCE 1982;108(9):959–77.
7. Ellingwood B, MacGregor J, Galambos TV, Cornell CA. Probability based load criteria: load factors and load combinations. J StructDiv, Reliability Design Methodology for Confined High Pressure... 155 ASCE 1982;108(9):978–97.
8. Ang AH, Cornell CA. Reliability bases of structural safety and design. J StructDiv, ASCE 1974;100(9):1755–69.
9. Lloyd G. Design and construction of large modern yacht rigs; 2009. <http://www.gl-group.com/infoServices/rules/pdfs/english/glrp-e.pdf>.
10. Tadich JK, Wedel-Heinen J, Petersen P. New guidance for the development of wind turbine blades; 2005. <http://wind.nrel.gov/public/SeaCon/Proceedings/Copenhagen.Offshore.Wind.2005/documents/papers/Large_offshore_wind_turbines/josef_kryger_New_guidance_for_the_developmentofwindturb.pdf>.
11. Veritas DN. Structural reliability analysis of marine structures; 1992. <http://exchange.dnv.com/Publishing/CN/CN30-6.pdf>.
12. NRCC, National Building Code of Canada, National Research Council of Canada; 2005. <http://www.nrc-cnrc.gc.ca/eng/ibp/irc/codes/05-national-buildingcode.html>.
13. Salmon CG, Johnson JE, Malhas FA. Steel structures – design and behavior. 5th ed. Upper Saddle River (NJ): Pearson-Prentice Hall; 2009.
14. Barbero EJ, Gutierrez JM. Determination of basis values from experimental data for fabrics and composites. In: SAMPE 2012, Baltimore, Maryland; May 21–24, 2012. p. 1–11.
15. Barbero EJ. Introduction to composite materials design, 2nd ed., 1st ed. Philadelphia (PA): CRC Press; 2010.

16. Lawless JF. Statistical models and methods for lifetime data. New York: John Wiley and Sons; 1982.
17. Ravindra MK, Galambos TV. Load and resistance factor design for steel. J StructDiv, ASCE 1978;9:1337–53.
18. Pai PF. Highly flexible strucutres: modeling, computation and experimentation. AIAA American Institute of Aeronautics and Astronautics; 2007.
19. Molina JC. Mechanical characterization of fabrics for inflatable structures. Master's thesis. West Virginia University, Morgantown, WV; 2008.
20. Weadon TL. Long term loading and additional material properties of Vectran fabric for inflatable structure applications. Master's thesis. West Virginia University, Morgantown, WV; 2010.
21. Reinhardt HW. On the biaxial testing and strength of coated fabrics. ExpMech 1976;16(2):71–4.
22. ASTM D 5035-06. Standard test method for breaking force and elongation of textile fabrics (strip method); 2006.
23. Rabinowicz E. Friction and wear of materials. New York: Wiley; 1995.
24. Rackwitz R, Flessler B. Structural reliability under combined random load sequences. ComputStruct 1978;9:484–94.
25. ASCE 7-5. Minimum design loads for buildings and other structures. American Society of Civil Engineers; 2006.
26. Dhillon BS. Mechanical reliability: theory models and applications (AIAA education series). Amer. Inst. of Aeronautics and Astronautics; 1988.
27. Kapur KC, Lamberson LR. Reliability in engineering design. 1st ed. Wiley and Sons; 1977.
28. Madsen HO, Krenk S, Lind NC. Methods of structural safety. Prentice-Hall; 1986.

CITATION

E.J. Barbero, E.M. Sosa, X. Martinez, J.M. Gutierrez, Reliability design methodology for confined high pressure inflatable structures, Engineering Structures, Volume 51, June 2013, Pages 1-9, ISSN 0141-0296, http://dx.doi.org/10.1016/j.engstruct.2013.01.011.

Chapter 4

Effect of Nitrite Ions on Steel Corrosion Induced by Chloride or Sulfate Ions

Zhonglu Cao, Makoto Hibino, and Hiroki Goda

Concrete Laboratory, Department of Civil Engineering, Kyushu Institute of Technology, Kitakyushu-shi 804-8550, Japan

ABSTRACT

The influence of nitrite concentration on the corrosion of steel immersed in three simulated pH environments containing chloride ions or sulfate ions has been investigated by comparing and analyzing the change of half-cell potential, the change of threshold level of Cl⁻ or So₄⁻, the change of threshold level of NO_2^-/Cl^- or NO_2^-/SO_4^{2-} mole ratio, and the changes of anodic/cathodic polarization curves and Stern-Geary constantB. The corrosivity of chloride ions against sulfate ions also has been discussed in pH 12.6, pH 10.3, and pH 8.1 environments containing 0, 0.053, and 0.2 mol/L NO_2, respectively.

INTRODUCTION

The corrosion of reinforcing steel in concrete has become one of the most severe deterioration mechanisms in concrete structures. It is generally accepted that due to the high alkalinity of cement hydration products, a protective layer of iron oxides is formed on the surface of steel, which provides adequate corrosion resistance. However, with the penetration of chloride, sulfate, and carbon dioxide and the appearance of concrete cracking, this protective layer becomes unstable and corrosion initials.

As one of the methods is to prevent steel corrosion, nitrite-based corrosion inhibitors, irrespective of being directly added into concrete during the mixing process or penetrating into concrete by the surface-applied remedial treatment, have been widely investigated in chloride-contaminated concrete [1–3], carbonated concrete [4–6], and cracked concrete [7, 8], and their inhibiting efficiencies also have been checked in simulated concrete pore solution, such as in highly alkaline solution [9–12], carbonated solution [12–14], and neutral and acid solution [12, 15]. Most of these results confirm the effectiveness of nitrite in increasing the chloride threshold level, delaying the onset of corrosion, and reducing the corrosion rate once the corrosion was initiated, but there is no general consensus on the NO_2^-/Cl^- mole ratio above that the preservation of the passive state can be ensured; suggested values for this threshold range from 0.11 to 1.0 in concrete and from 0.07 to greater than 2 in simulated pore solution. This difference in the threshold level of NO_2^-/Cl^- mole ratio obtained from various literatures might be due to the way of determining the concentrations of chloride and nitrite in concrete (free ions and total ions, etc.), the different qualities of mortar and concrete used in the experiments, the different components of simulated pore solution, and the different surface topographies and compositions of the steel.

Comparing and analyzing these literatures, the authors find the contents of available nitrite are different in these studies. So whether it is possible that the nitrite concentration has an influence on this threshold value attracts the interest of us, and up to now, no published literature gives the answer.

Another problem here is that the inhibiting efficiency of nitrite in carbonated concrete with or without chloride is not ideal, so whether the high pH environment plays a role in assisting nitrite to inhibit chloride-induced corrosion, as far as known by the authors, no literature clarifies this.

As described above, many researches have been done for chloride-induced steel corrosion. However, fewer works are focused on sulfate-induced steel corrosion, even the corrosion mechanism of sulfate ions is still not clear, and there are also some controversies on the corrosivity of sulfate against chloride ions in alkaline and neutral environments [16–23]. Additionally, all of these works for sulfate-inducedcorrosion are just performed in concrete or solution without the addition of any corrosion inhibitor. Little attention has been given to the effect of nitrite ions on sulfate-induced corrosion, and furthermore, reports on the corrosivity of sulfate against chloride, under the action of nitrite, are extremely scarce.

Besides these, the effect of nitrite ions on the polarization behaviors of steel also has attracted the interest of the authors. As it is known that the corrosion current density of steel is usually calculated by the use of the Stern-Geary equation defined as $i_{corr} = B/Rp$, where Rp is the polarization resistance and B is a constant which varies with the expression that $B = \beta_a \beta_c / (2.3(\beta_a + \beta_c))$, where βa and βc are the anodic and cathodic Tafel slopes obtained from the anodic and cathodic polarization curves of steel, respectively. Generally, the calculation of i_{corr} mainly focuses on the determination of Rp and rarely gives an attention to the value of B. The value of B is commonly considered to be 26 mV for steel in corroded state and 52 mV for steel in passive state [24]. As we know, the polarization behavior of steel is affected by many factors, such as oxygen and corrosion inhibitors, so the direct use of these B values is suitable or not, is worthy of discussing. What is more, with the increasing application of corrosion inhibitors, some concern should be given to the effect of corrosion inhibitors on B value. Nitrite that performs as an anodic inhibitor can repair the passive film and improve the polarization resistance Rp; however, whether it has an effect on Stern-Geary constant B should be checked and clarified, because in most cases, especially for onsite measurement of corrosion rate, the value of B is taken for granted without any experimental verification, so the calculation of corrosion current density might be misled if any influence of nitrite on the Tafel slope exists.

For all these reasons, the main objective of this paper was to investigate and clarify the influence of nitrite concentration and pH environments on the inhibiting efficiency of nitrite and the influence of nitrite on the anodic/cathodic polarization curves of steel and Stern-Geary constant B. Additionally, the corrosivity of chloride and sulfate on steel was also compared. Because the corrosion of steel in concrete is difficult to be investigated, and so all experiments in this study were carried out in simulated chemical environments.

EXPERIMENTAL

Cold-rolled carbon steel sheets (JIS G3141 SPCC-SB) with a dimension of 60 × 60 × 1 mm were selected to use in this study. The composition of the steel sheet was (wt%): 0.1025% C, 0.5204% Mn, 0.0193% P, 0.0097% S, and balance Fe. At one corner, lead wire was fixed to the steel by screw and the connection area was sealed by epoxy resin. The exposed area of the steel was 67.7 cm². Before the experiments, steel sheets were polished by sandpaper and cleaned with acetone. The corrosion of steel sheets was investigated in three different pH environments (pH 12.6, pH 10.3, and distilled water) simulating the highly alkaline environment, weakly alkaline environment, and neutral environment that really existed on the surface of steel in concrete. The pH 12.6 and pH10.3 environments were made by mixing different contents of LiOH and H_3BO_3. LiOH and H_3BO_3 were chosen for the following reasons. Firstly, they worked as buffer solution which could make the pH in a constant level during the whole experiment process, and secondly LiOH was often used as electrolyte for electrochemical removal of chloride from concrete structures, and thirdly Li^+ that is mainly in the form of $LiNO_2$ was usually added into concrete to prevent the concrete expansion caused by alkali-silica reaction. The use of LiOH and H3BO3 could not reflect the real concrete pore solution that is composed of Na^+, K^+, Ca^{2+}, $OH-$, CO_3^{2-}, HCO_3^-, and so forth, but it still can simulate the possible pH environments that exist in concrete. Although the pore solution compositions (cation type and anion type) had an effect on the protective properties of the passive oxide films and the corrosion behavior of steel [6, 13, 25–27], the focus of this work was to investigate the influence of pH environments and nitrite ions on steel corrosion induced by chloride or sulfate, not to study the effect of other cation/anion ions. Therefore, the use of LiOH and H_3BO_3 might be not the best, but it was suitable in this work.

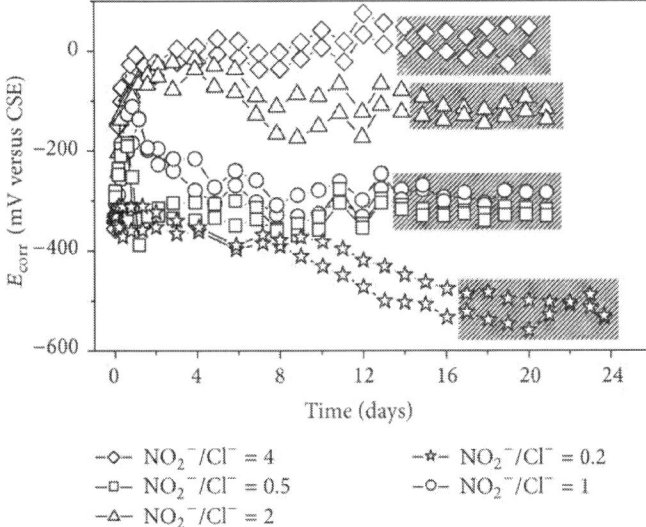

Figure 1: Time evolution curves of E_{corr} in distilled water containing 0.2 mol/L NO2− and with different mole ratio of NO^{2-}/Cl^-.

For each pH environment, sodium nitrite was added with the concentration of 0, 0.02, 0.053, 0.2, 0.53 mol/L, respectively, while sodium chloride or sodium sulfate was added according to the mole ratio of NO^{2-}/Cl^- or $NO2-/SO4^{2-}$ given in Table 1 or Table 2, respectively. With the presence of nitrite, the pH of distilled water increased to about 8.1

Table 1: Experimental design for chloride-induced corrosion.

pH	NO_2^- mol/L	Cl^- mol/L							
pH 12.6 ± 0.3	0	—	—	0.4	0.2	0.1*	0.05	—	0*
	0.02	—	0.1*	—	—	—	—	—	—
	0.053	0.53	0.265	0.106	0.053	0.027	—	—	0
	0.2	—	1	0.4	0.2	0.1*	0.05	—	0*
	0.53	—	—	1.06	0.53	0.265	0.133	0.066	0
pH 10.3 ± 0.3	0	—	—	0.4	0.2	0.1*	0.05	—	0*
	0.02	—	0.1*	—	—	—	—	—	0*
	0.053	0.53	0.265	0.106	0.053	0.027	—	—	0
	0.2	—	1	0.4	0.2	0.1*	0.05	—	0*
	0.53	—	—	1.06	0.53	0.265	0.133	0.066	0
Distilled water	0	—	—	0.4	0.2	0.1*	0.05	—	0*
	0.02	—	0.1*	—	—	—	—	—	0*
	0.053	0.53	0.265	0.106	0.053	0.027	—	—	0
	0.2	—	1	0.4	0.2	0.1*	0.05	—	0*
	0.53	—	—	1.06	0.53	0.265	0.133	0.066	0
Mole ratio of NO_2^-/Cl^-		0.1	0.2	0.5	1	2	4	8	—

Table 2: Experimental design for sulfate-induced corrosion.

pH	NO_2^- mol/L	SO_4^{2-} mol/L						
pH 12.6 ± 0.3	0	—	—	—	0.2	0.1*	0.05	0.025
	0.02	0.2	0.1*	0.04	0.02	0.01	—	—
	0.053	—	0.265	0.106	0.053	0.027	0.013	—
	0.2	—	—	0.4	0.2	0.1*	0.05	0.025
pH 10.3 ± 0.3	0	—	—	—	0.2	0.1*	0.05	0.025
	0.02	0.2	0.1*	0.04	0.02	0.01	—	—
	0.053	—	0.265	0.106	0.053	0.027	0.013	—
	0.2	—	—	0.4	0.2	0.1*	0.05	0.025
Distilled water	0	—	—	—	0.2	0.1*	0.05	0.025
	0.02	0.2	0.1*	0.04	0.02	0.01	—	—
	0.053	—	0.265	0.106	0.053	0.027	0.013	—
	0.2	—	—	0.4	0.2	0.1*	0.05	0.025
Mole ratio of NO_2^-/SO_4^{2-}		0.1	0.2	0.5	1	2	4	8

For each combination of NO^{2-} and Cl^- or each combination of NO_2^- and SO_4^{2-} in three different pH environments as given in Table 1 or Table 2, two specimens were performed at the same time, and half-cell potential referred to as $Cu/CuSO_4$ Electrode (CSE) was measured at set intervals until the potential reached a constant value. The time taken for the stable of potential was more than 10 days as shown in Figure 1. The final half-cell potential of steel, in each given combination, was calculated by averaging the potential value of two specimens, where the value of each specimen wasdetermined by averaging the data obtained in the stable part of potential versus time curve (as shaded part shown in Figure 1). During the whole experimental process, the pH value and temperature of solution were monitored all the time to make sure that they still maintained a constant level, and the time taken for the visible formation of rust was alsorecorded. At the end of the experiments, specimens weretaken out from the solutions andexamined the formationof any visual rust on their surface. Whether the corrosionoccurred or not for each combination was judged according to Table 3.

Table 3: Corrosion judgment defined in this study

Case	Specimen 1	Specimen 2	Judgments	Marked as
1	Corrosion	Corrosion	Corrosion	◆ ▲ ● ■
2	Corrosion	No corrosion	Uncertain	◈ ▲ ◉ ▆
	No corrosion	Corrosion	Uncertain	
3	No corrosion	No corrosion	No corrosion	◇ △ ○ □

Anodic and cathodic polarization curves of steel were only carried out in the combinations marked as "*" in Tables 1 and 2. The reference electrode used here was Ag/AgClelectrode and the counter electrode was platinum. Prior to the polarization measurement, the steel had been immersed in the solution more than 15 days and the E_{corr} had reacheda stable value. The cathodic polarization was done firstly, which started from E_{corr} toward the negative direction with the interval of −25 mV (at the changing rate of 1 mV/s). After the potential was set, 30 seconds were necessary to wait before writing down the current value. After finishing cathodic polarization, the steel was continued to be immersed in the solution about one day for the recovery of Ecorr. And after that, the anodic polarization was started from Ecorr toward thepositive direction with the interval of 25 mV (at the speed of1 mV/s). After setting the potential, 30 seconds are necessaryto wait before writing down the current value. Generally,a potential scan greater than ±(50 to 100) mV about $E_{corr}$is required to reach the potential at which the anode Tafel or cathode Tafel behavior dominates and linear polarization is expected [28]. Tafel slopes β_a and β_c in this study were calculated by using the data that obtained between E_{corr} ± 75 and E_{corr} ± 300.

RESULTS AND DISCUSSION

According to the experimental results, the Ecorr of steel immersed in different pH environments with the absence and presence of NO_2^-/Cl^- or NO_2^-/SO_4^{2-}, as given in Tables 1 and 2, was shown in Figures 2 and 3. The Ecorr presented here was calculated by averaging the data obtained in the stable part of potential versus time curve. The surface state of steel was marked as filled circle for corrosion and empty circle for no corrosion. From the figure, it can be found thatthere was a pretty clear dividing line between passive state and corrosion state. The potential that changes abruptly from passive state to corrosion state is called Flade potential (EF), which is the lowest potential of film formation and has a good linear relationship with pH values as reported by other literatures [29, 30]. Usually, the lower the EF is, the stronger the passivation ability of steel is, and the passive film becomes more stable.

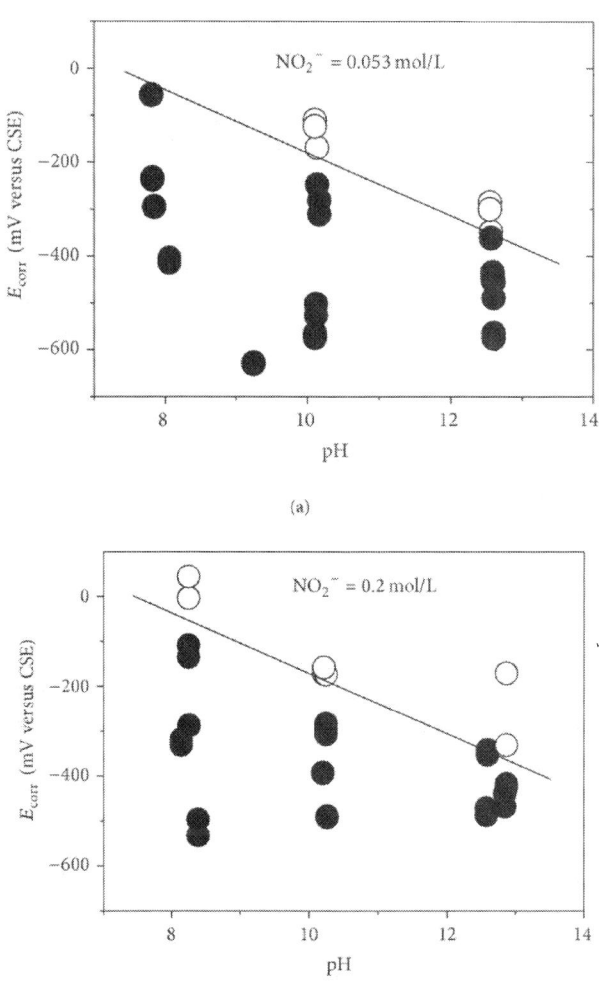

110 | RESULTS AND DISCUSSION

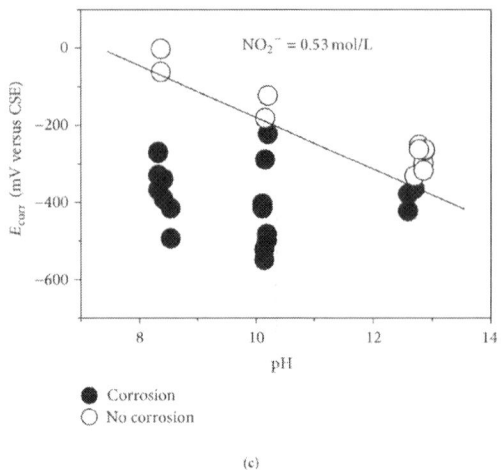

(c)

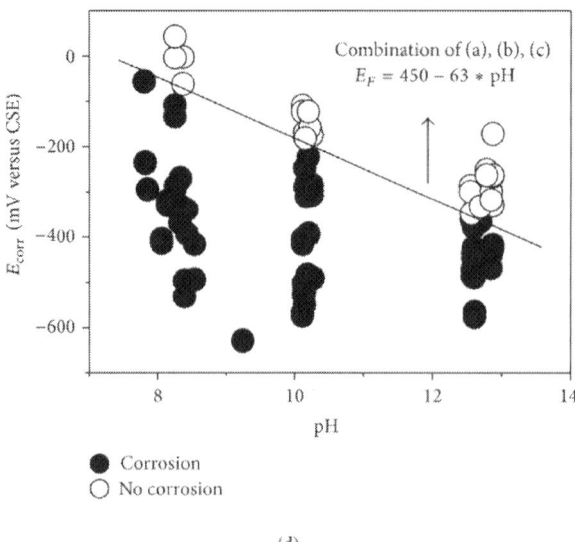

(d)

Figure 2: Relationship between EF and pH for chloride-induced corrosion. (a) NO_2^- = 0.053 mol/L, (b) NO_2^- = 0.2 mol/L, (c) NO_2^- =0.53 mol/L, and (d) combination of (a), (b), and (c).

The possible influence of NO_2^- concentration on the relationships between EF and pH for chloride-induced corrosionand sulfate-induced corrosion also had been investigated. From Figures 2 and 3, it can be found that the NO_2^- cocentration that changed from 0.053 mol/L to 0.53 mol/L for chloride-induced corrosionor from 0.02 mol/L to 0.2 mol/L for sulfate-induced corrosion had little influence on the linear relationship between EF and pH, which also means that the use of the relationship between EF and pH as the criterion for corrosion judgment is reasonable in this work. So, in this study, based on the experimental results, the relationships between EF and pH were given out in the form of $EF = 450 - 63 * pH$ for chloride-induced corrosion as shown in Figure 2(d) and $EF = 300 - 50 * pH$ for sulfate-induced corrosion as given in Figure 3(d), which were used as the corrosion criterion for the following result analysis.

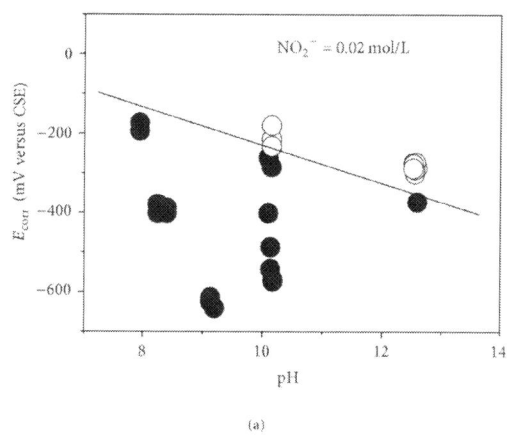

(a)

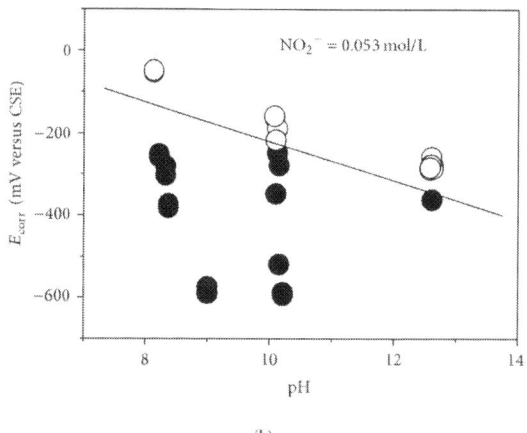

(b)

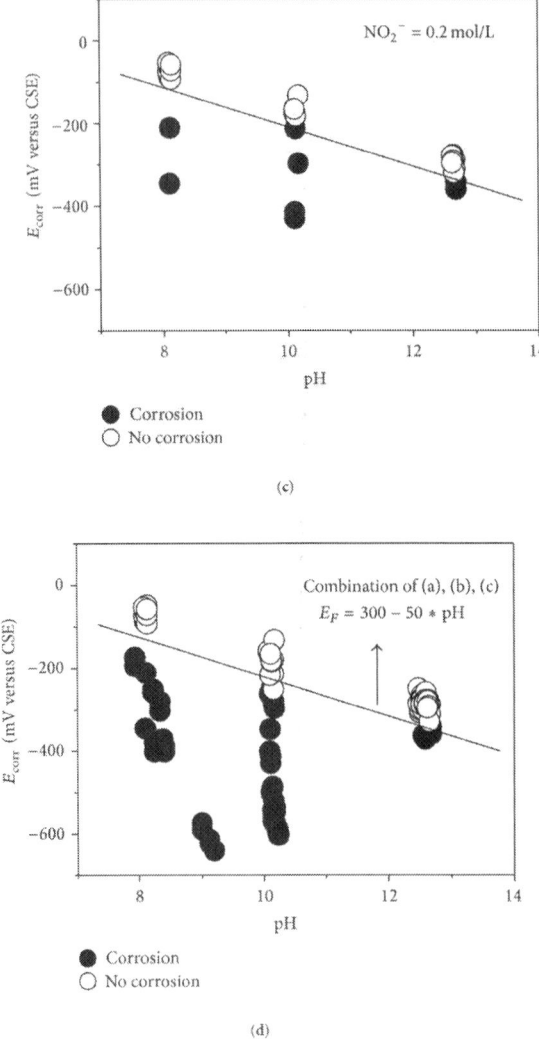

Figure 3: Relationship between EF and pH for sulfate-induced corrosion. (a) NO_2^- = 0.02 mol/L, (b) NO_2^- = 0.053 mol/L, (c) NO_2^- =0.2 mol/L, and (d) combination of (a), (b), and (c).

E_{corr} Induced by Cl^-

Figure 4 describes the effect of nitrite concentration on E_{corr} induced by chloride in three pH environments. The E_{corr} of steel obviously increased withthe presence of nitrite for all pH environments, especially for low pH environment. The increment of potential in pH 8.1 was greater than that in pH 12.6. So with the decreasing of pH, nitrite became more effective to increase the E_{corr}, but it was not helpful to inhibit the corrosion. This was because the EF was −344 mV for pH 12.6, −199 mV for pH 10.3, and −60 mV for pH 8.1, increased with the decreasing of pH. If the E_{corr} was greater than EF, the steel would be passivated; otherwise, the steel was corroded.

Nitrite concentration and pH had an influence on the increment of E_{corr}. In pH 12.6 environment, for the given chloride content, the E_{corr} of steel increased with the increasing of nitrite concentration, and this increment was moreevident when chloride content was higher. However, in pH10.3 and pH 8.1 environments, when nitrite was increased from 0.2 to 0.53 mol/L, the increment of steel potential were not obvious. The E_{corr} with 0.53 mol/L nitrite was equal to or even lower than that with 0.2 mol/L nitrite. No benefiton the further increase of E_{corr} can be obtained from higher nitrite concentration in pH 10.3 and pH 8.1 environments, and this indicates that the ability of nitrite to improve steel potential was limited and weakened.

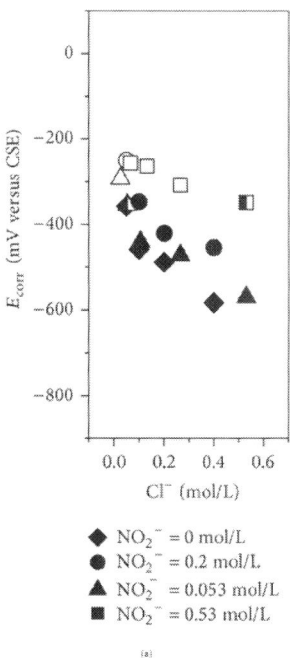

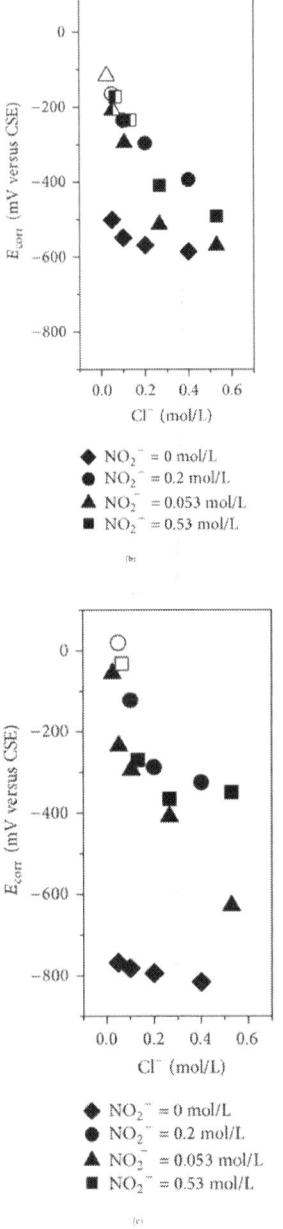

Figure 4: Effect of NO_2^- and pH on E_{corr} induced by Cl–. (a) pH = 12.6, (b) pH = 10.3, and (c) pH = 8.1.

E_{corr} Induced by SO_4^{2-}

Figure 5 presents the effect of nitrite on E_{corr} induced by sulfate in three pH environments. In pH 12.6, irrespective of the presence or absence of nitrite, E_{corr} did not change significantly with the increasing of sulfate, even no rust can be found on the surface of steel after 23 days. The increment of E_{corr} was negligible after the presence of nitrite. In pH 10.3, E_{corr} decreased with the increasing of sulfate concentration. Rust can be found on the surface of steel, but most of the area was still in good state. In pH 8.1, with the absence of nitrite, the entire steel surface was corroded severely. E_{corr} increased with the presence of nitrite, but low concentration of nitrite (0.02 mol/L) resulted in the formation of corrosion pitting on the surface of steel. The concentrations of nitrite had an influence on steel potential. For highly alkaline environment with pH 12.6, with the increasing of nitrite concentration, the increment of potential was not obvious, even the decrease of potential can be found according to the experimental results. However, for the conditions of pH 10.3 and pH 8.1, with the raising of nitrite concentration, the potential increased significantly, and this increment was more evident when sulfate content was high.

Relationship between E_{corr} and NO_2^-/Cl^-

E_{corr} had a good linear relationship with $\log(NO_2^-/Cl^-)$ and increased with the increasing of NO_2^-/Cl^- mole ratio, as shown in Figure 6. In pH 12.6 environment, the concentration of nitrite almost had no influence on the curve of E_{corr} versus $\log(NO2^-/Cl^-)$, but in pH 10.3 and pH 8.1 environments, with the increasing of nitrite concentration, the curve shifted toward the direction of higher $NO2^-/Cl^-$ mole ratio, which resulted in the decrease of E_{corr} for the same NO_2^-/Cl^-. This means that the inhibiting ability of nitrite to per unit of chloride was weakened in pH 10.3 and pH 8.1 environments compared to that in pH 12.6 environment.

For the specified nitrite concentration at constant pH, the changing of NO_2^-/Cl^- mole ratio not only affected the E_{corr}, but also influenced the initial time of corrosion. For example, in distilled water with 0.2 mol/L nitrite (as shown in Figure 1), when the mole ratio of NO_2^-/Cl^- was changed with 4, 2, 1, 0.5, and 0.2, the stable self-potential was changed correspondingly with 20, −121, −288, −325, and −514 mV, and the initial time of corrosion was also changed correspondingly with >20, 5.8, 1.6, 0.8, and 0.1 days. Both E_{corr} and the corrosion initial time increased with the increasing of NO_2^-/Cl^-, which can also be confirmed in other conditions shown in Table 1. All this indicates that the value of $NO2-/Cl^-$ had great influence on steel corrosion. The higher value of NO_2^-/Cl^- makes the steel passivate, while the lower value of NO_2^-/Cl^- makes the steel corrode. So the use of NO_2^-/Cl^- mole ratio as parameter to evaluate the inhibiting efficiency of nitrite on chloride-induced corrosion is feasible, and there must be a threshold level of NO_2^-/Cl^- which can determine whether the steel is corroded or not. In this study, the threshold levels of NO_2^-/Cl^- can be calculated based on the EF and the good linear relationship between E_{corr} and $\log(NO_2^-/Cl^-)$, as given in Table 4.

Table 4: Relationship between E_{corr} and $\log(NO^{2-}/Cl^-)$

NO_2^- mol/L	pH	E_F, mV versus CSE	$E_{corr} = a^*\log(NO_2^-/Cl^-) + b$		
			a	b	R^2
0.053	12.58	−342.5	201.9	−357.2	0.97
0.2	12.72	−351.4	172.6	−386.9	0.90
0.53	12.71	−350.7	124.3	−353.1	0.95
0.053	10.13	−188.2	370.2	−215.1	0.98
0.2	10.24	−195.1	253.0	−310.9	0.995
0.53	10.14	−188.8	326.5	−466.8	0.96
0.053	7.88	−46.4	398.0	−189.3	0.96
0.2	8.26	−70.4	396.7	−237.7	0.98
0.53	8.42	−80.5	307.4	−386.6	0.83

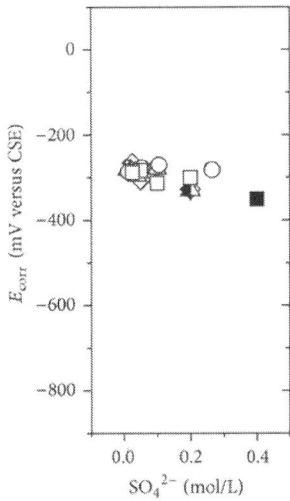

- ◆ $NO_2^- = 0$ mol/L
- ● $NO_2^- = 0.053$ mol/L
- ▲ $NO_2^- = 0.02$ mol/L
- ■ $NO_2^- = 0.2$ mol/L

(a)

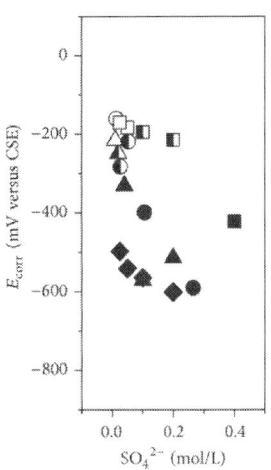

- ◆ $NO_2^- = 0$ mol/L
- ● $NO_2^- = 0.053$ mol/L
- ▲ $NO_2^- = 0.02$ mol/L
- ■ $NO_2^- = 0.2$ mol/L

(b)

118 | RESULTS AND DISCUSSION

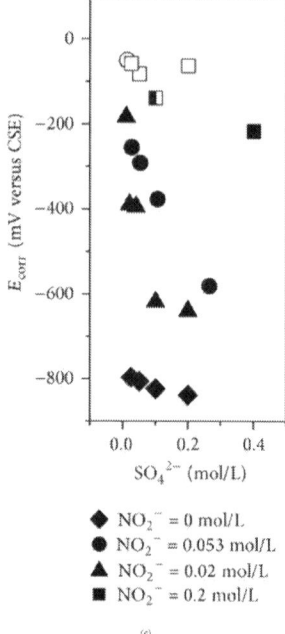

- ◆ $NO_2^- = 0$ mol/L
- ● $NO_2^- = 0.053$ mol/L
- ▲ $NO_2^- = 0.02$ mol/L
- ■ $NO_2^- = 0.2$ mol/L

(c)

Figure 5: Effect of NO_2^- and pH on E_{corr} induced by SO_4^{2-}. (a) pH = 12.6, (b) pH = 10.3, and (c) pH = 8.1.

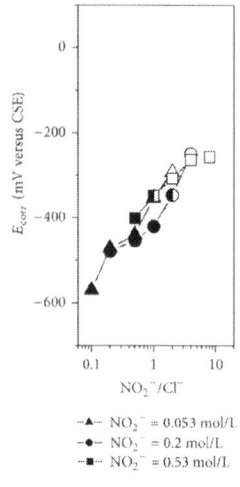

- ▲ $NO_2^- = 0.053$ mol/L
- ● $NO_2^- = 0.2$ mol/L
- ■ $NO_2^- = 0.53$ mol/L

(a)

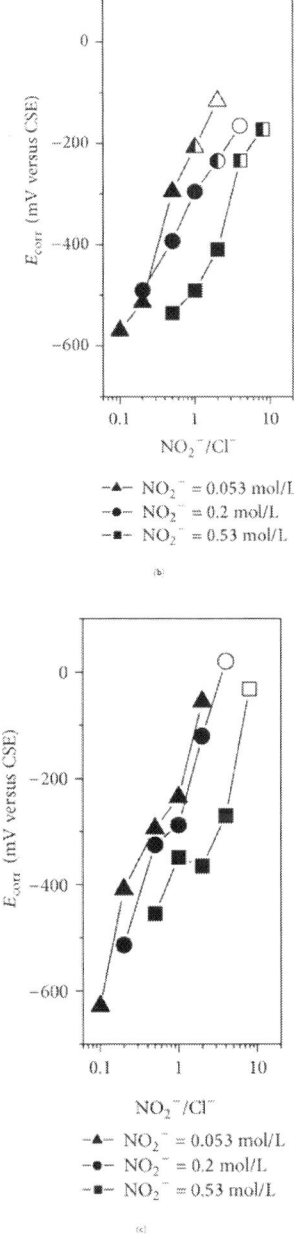

Figure 6: Effect of NO2⁻ on E_{corr} versus $\log(NO_2^-/Cl^-)$ curves. (a) pH = 12.6, (b) pH = 10.3, and (c) pH = 8.1.

Relationship between E_{corr} and NO_2^-/SO_4^{2-}

The relationship between E_{corr} and $\log(NO_2^-/SO_4^{2-})$ shown in Figure 7 indicates that, in pH 12.6 environment, the mole ratio of NO_2-/SO_4^{2-} had little influence on E_{corr}, while in pH 10.3 and pH 8.1 environments, E_{corr} increased with the increasing of NO_2-/SO_4^{2-} mole ratio. For sulfate-induced corrosion, the nitrite concentration almost had no influence on the curve of E_{corr} versus $\log(NO_2^-/SO_4^{2-})$, which was completely different from the condition for chlorideinduced corrosion. The linear relationship between E_{corr} and $\log(NO_2^-/SO_4^{2-})$ was shown in Table 5.

Table 5: Relationship between E_{corr} and $\log(NO_2^-/SO_4^{2-})$.

NO_2^- mol/L	pH	E_F, mV versus CSE	$E_{corr} = a*\log(NO_2^-/SO_4^{2-}) + b$		
			a	b	R^2
0.02	12.6	−330	27.9	−284.1	0.47
0.053	12.6	−330	−6.3	−279.8	0.25
0.2	12.6	−330	47.0	−321.4	0.69
0.02	10.1	−205	283.1	−280.2	0.87
0.053	10.1	−205	310.3	−324.3	0.85
0.2	10.1	−205	177.3	−290.4	0.65
0.02	8.26	−113	347.6	−327.5	0.93
0.053	8.26	−113	371.4	−303.7	0.96
0.2	8.1	−105	98.7	−141.4	0.50

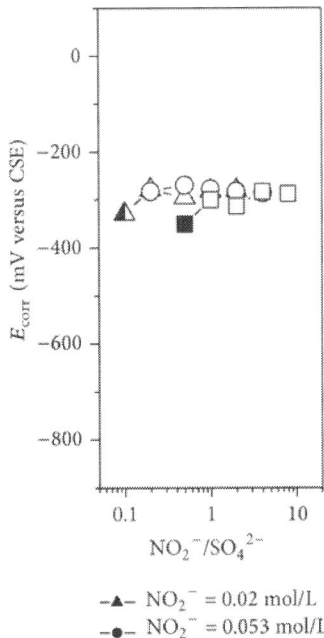

(a)

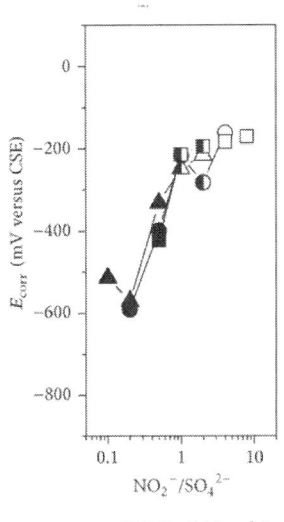

(b)

RESULTS AND DISCUSSION

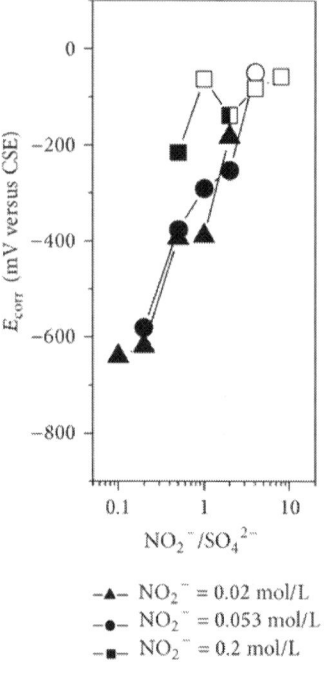

- ▲ NO_2^- = 0.02 mol/L
- ● NO_2^- = 0.053 mol/L
- ■ NO_2^- = 0.2 mol/L

(c)

Figure 7: Effect of NO_2^- on E_{corr} versus $\log(NO_2^-/SO_4^{2-})$ curves. (a) pH = 12.6, (b) pH = 10.3, (c) pH = 8.1.

Threshold Level of NO_2^-/Cl^-

Figure 8 presents the effect of nitrite concentration and pH on the threshold level of NO_2^-/Cl^- that was determined by using the EF and the relationship between Ecorr and $\log(NO_2^-/Cl^-)$ as shown in Table 4. In pH 12.6 environment, nitrite concentrationalmost had no influence on the threshold level of NO_2^-/Cl^-, while in pH 10.3 and pH 8.1 environments, the threshold level increased with the increasing of nitrite concentration. The constant of the threshold level of $NO2^-/Cl^-$ in pH 12.6 environment means that NO^{2-}/Cl^- mole ratio can be performed as the parameter to evaluate the corrosion of steel in highly alkaline environment, but it is not suitable for the situation in weakly alkaline and neutral environmentsbecause the threshold level of NO_2^-/Cl^- changed greatly with different nitrite concentrations.

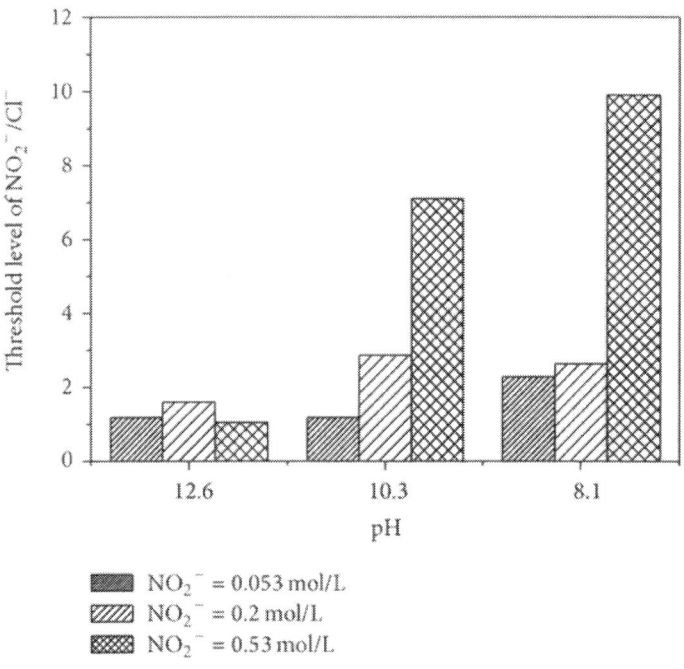

Figure 8: Effect of NO_2^- on the threshold values of NO_2^-/Cl^-.

Threshold Level of NO_2^-/SO_4^{2-}

In the NO_2^-/Cl^- system, there was a good linear relationship between E_{corr} and $\log(NO_2^-/Cl_-)$ for each nitrite concentration in pH 12.6, pH 10.3, and pH 8.1 environments, but in NO2–/SO_4^{2-} system, the good linear relationship between E_{corr} and $\log(NO^{2-}/SO_4^{2-})$ only can be found in pH 8.1 and pH10.3 environments with 0.02 and 0.053 mol/L NO_2^-. So the threshold level of NO_2^-/SO_4^{2-} in pH 8.1 and pH 10.3 environments with 0.02 and 0.053 mol/L NO_2^- was obtained by using the EF and the relationship between E_{corr} and $\log(NO_2^-/SO_4^{2-})$ as shown in Table 5, while the threshold levels of NO^{2-}/SO_4^{2-} in pH 12.6 environment with 0.02, 0.053, and 0.2 mol/L NO_2^- and in pH 8.1 and pH 10.3 environments with 0.2 mol/L NO^{2-} were judged according to the results of corrosion judgment as described in Table 3 and shown in Figure 7. The threshold levels of NO_2^-/SO_4^{2-} obtained here might be not accurate, but they were well consistent with the experimental results and could provide some information on the influence of nitrite concentrations.

Figure 9 presents the effect of nitrite concentrations and pH values on the threshold levels of NO_2^-/SO_4^{2-}. It can be seen that the threshold value of NO_2^-/SO_4^{2-} was greatly influenced by pH values and nitrite concentrations and decreased with the increasing of pH values. In pH 8.1 environments, this threshold value decreased with the increasing of nitrite content, while in pH 10.3 environments, it showed the opposite trend, increased with the increasing of nitrite content. In pH 12.6 environments, the threshold value of NO_2^-/SO_4^{2-} was 0.2, 0.2, and 0.656 for 0.02, 0.053, and 0.2 mol/L nitrite, respectively. The change of the threshold value in these pH environments means that it cannot be used as the parameter to evaluate the inhibiting efficiency of nitrite on sulfate-induced corrosion.

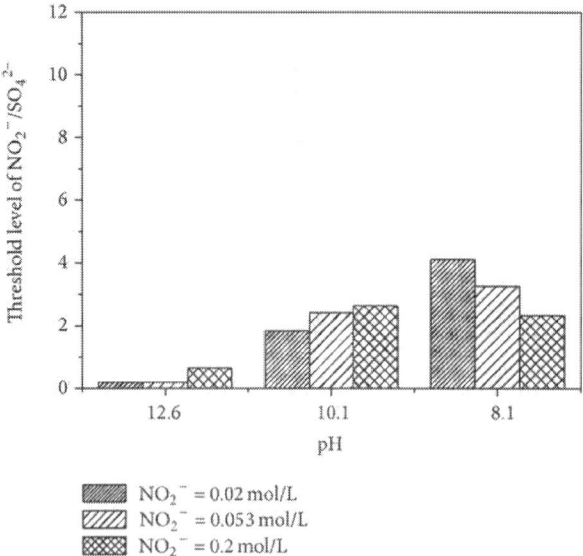

Figure 9: Effect of NO_2^- on the threshold values of $NO2-/SO_4^{2-}$.

Threshold Level of Cl⁻

The effect of nitrite concentration and pH on the threshold level of chloride is shown in Figure 10. In pH 12.6 environment, the chloride threshold level increased markedly when the concentration of nitrite was raised from 0 mol/L to 0.53 mol/L, while in pH 10.3 and pH 8.1 environments, if the concentration of nitrite was less than 0.2 mol/L, the chloride threshold level had an increasing trend, and if the nitrite content was higher than 0.2 mol/L, there was almost no increase in the chloride threshold level, even the decrease could also be found with the experimental results. The chloride threshold level in pH 12.6environmentwas greater than that in pH 10.3 and pH 8.1 environments, especially for the presence of higher nitrite concentration, indicating that the highly alkaline environment played an important role in assisting nitrite to increase the chloride threshold level. The combined effect of OH^- and NO_2^- oninhibiting corrosion was superior to the individual effect. However, the use of NO_2^-/Cl^- mole ratio as the parameter to guarantee the inhibition efficiency did not take into account the effect of OH^-. So whether NO_2^-/Cl^- is the most suitable parameter to evaluate chloride-induced corrosionin highly alkaline environments is worthy of further study. Besides this, why the inhibiting efficiency of nitrite on steel corrosion in neutral environment is weakened should be further investigated. The results obtained in this study are mainly based on the methods of half-cell potential and visual examination; they should be further confirmed by corrosion current measurement.

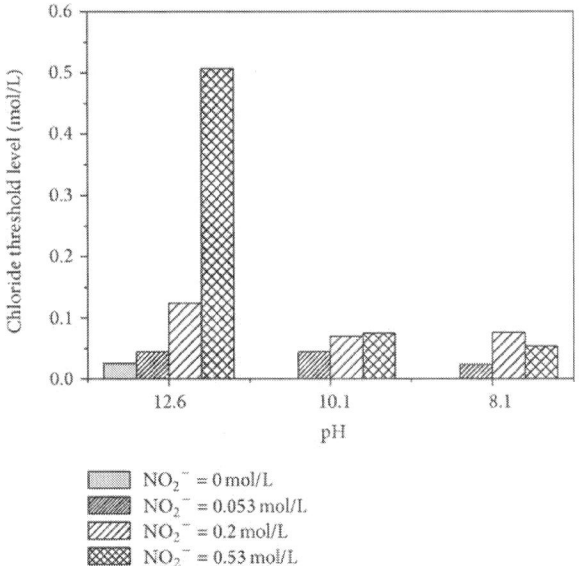

Figure 10: Effect of $NO2^-$ on the threshold values of Cl^-.

Threshold Level of SO_4^{2-}

The effect of nitrite concentration and pH on the threshold level of sulfate was given in Figure 11. As Seen from the figure, irrespective of the pH values, with the increasing of nitrite concentration, the sulfate threshold level increased significantly. For the same nitrite content, sulfate threshold level in pH 12.6 environments was much greater than that in pH 10.3 and pH 8.1 environments. So the highly alkaline environment played an important role in assisting nitrite to increase the sulfate threshold level.

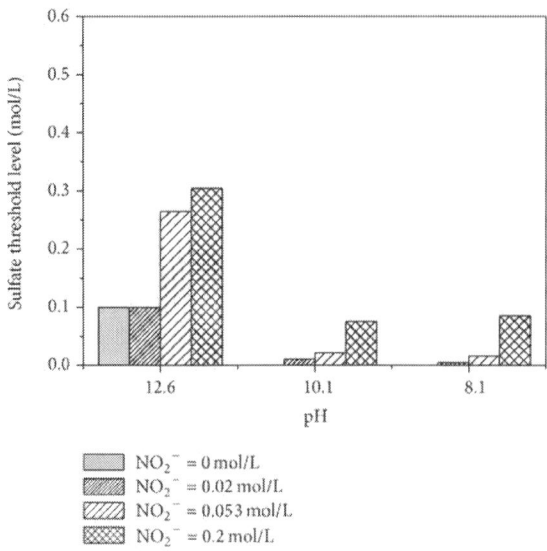

Figure 11: Effect of NO_2^- on the threshold values of SO_4^{2-}.

Comparing the Ecorr Induced by Cl^- and SO_4^{2-}

Figure 12 compares the E_{corr} induced by Cl^- and SO_4^{2-} in three pH environments containing 0, 0.053, and 0.2 mol/L NO_2^-. It can be seen that, in pH 10.3 and pH 8.1 environments with 0 and 0.053 mol/L NO_2^-, when SO_{42-} content was equal to Cl– content, the E_{corr} resulted from sulfate-induced corrosion was lower than that resulted from chloride-induced corrosion. The corrosivity of sulfate was stronger than that of chloride in this condition. However, there was a dramatic change after the content of nitrite was raised to 0.2 mol/L NO_2^-; the E_{corr} in sulfate environment was higher than thatin chloride environment and the corrosion extent of steel immersed in sulfate environment was greatly reduced. In pH 12.6 environment, regardless of the absence or the presence of nitrite, the E_{corr} resulted from sulfate-induced corrosion was higher than that resulted from chloride-induced corrosion, and almost all specimens were in passivation state when sulfate ions were equal to or less than 0.2 mol/L. So the corrosivity of sulfate ions on steel in pH 12.6 environment was weaker than that of chloride. From the above analysis, a conclusion could be obtained that the corrosivity of sulfate against chloride was variable and greatly affected by the value of pH and the concentrations of nitrite.

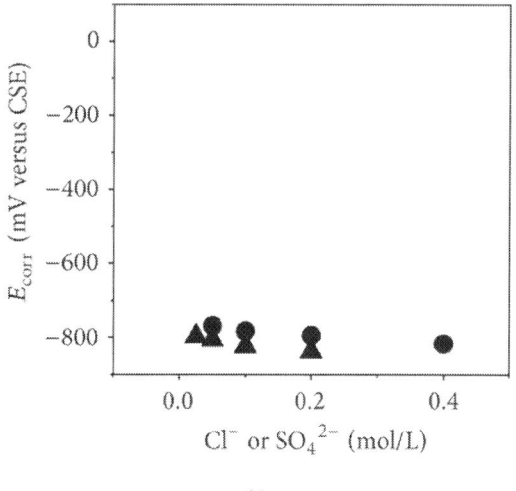

(a)

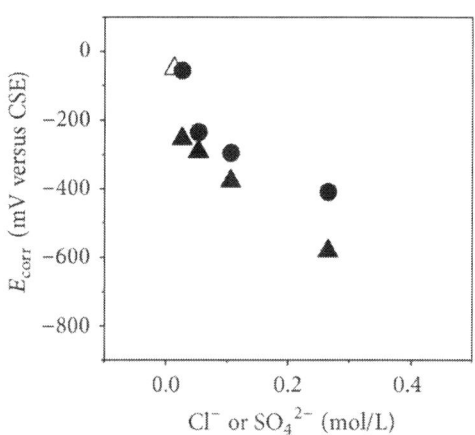

(b)

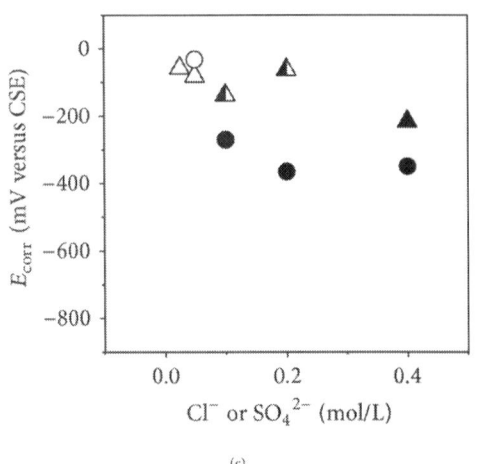

(c)

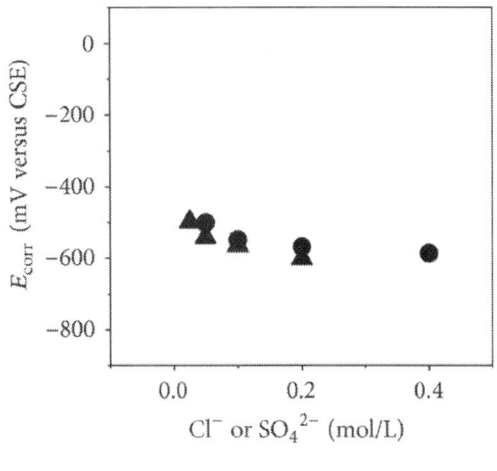

(d)

Guide to Stability Design Criteria for Metal Structures | 129

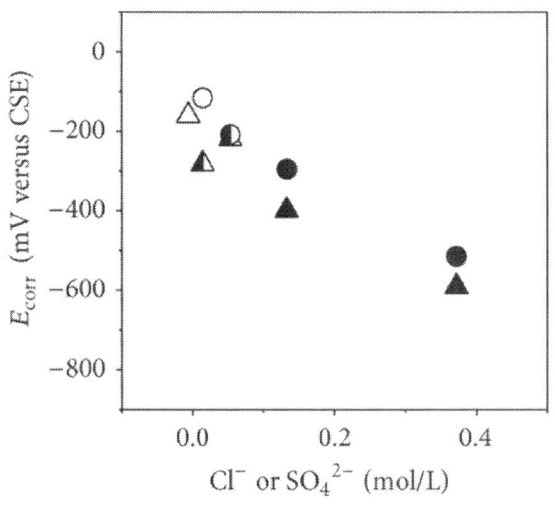

(e)

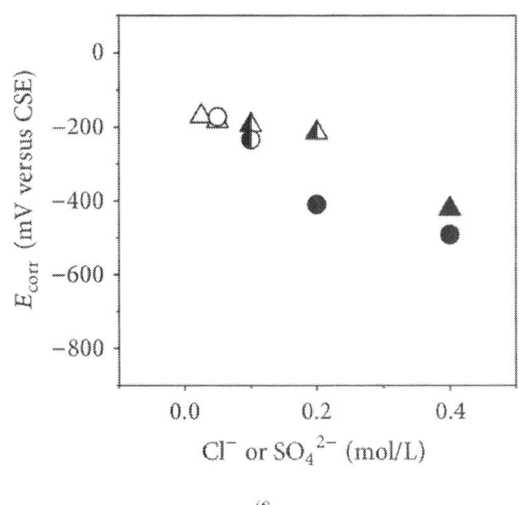

(f)

RESULTS AND DISCUSSION

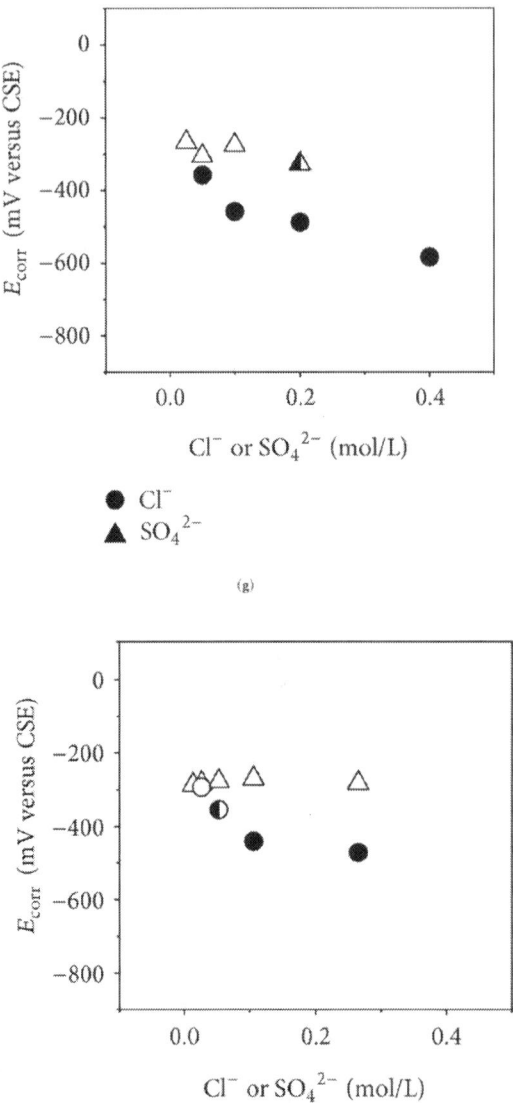

(g)

(h)

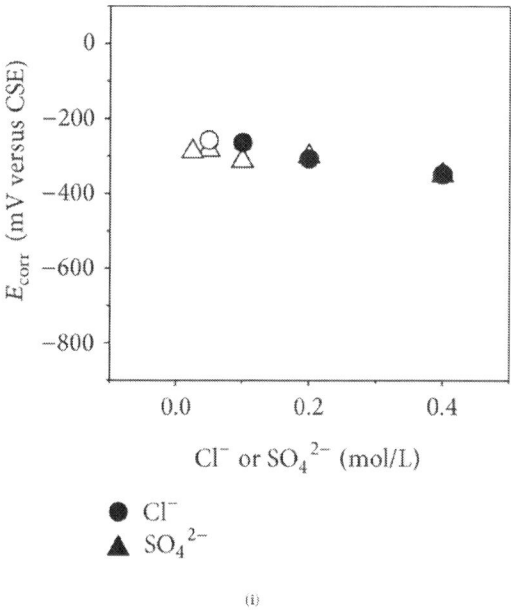

Figure 12: Comparing the E_{corr} induced by Cl^- and SO_4^{2-}. (a) pH 8.1, NO_2^- = 0 mol/L, (b) pH 8.1, NO_2^- = 0.053 mol/L, (c) pH 8.1, NO2− =0.2 mol/L, (d) pH 10.3, NO_2^- = 0.0 mol/L, (e) pH 10.3, NO_2^- = 0.053 mol/L, (f) pH 10.3, NO_2^- = 0.2 mol/L, (g) pH 12.6, NO_2^- = 0.0 mol/L, (h) pH 12.6, NO_2^- = 0.053 mol/L, (i) pH 12.6, NO_2^- = 0.2 mol/L.

Comparing the Threshold Level of Cl- and SO_4^{2-}

The threshold level of Cl^- and the threshold level of SO42−were compared and analyzed in pH 12.6, pH 10.3 and pH8.1 environments containing 0, 0.053, and 0.2 mol/L NO_2^- respectively. The results were presented in Figure 13. In pH 12.6 environments, whatever the concentration of nitrite was, sulfate threshold levelwas always higher than chloride threshold level under the same condition. This means that chloride-induced corrosion was more prone to initiate than sulfate-induced corrosion in highly alkaline environment. However, in pH 10.3 and pH 8.1 environments, with the presence of 0.053 mol/L nitrite, the sulfate threshold levelwas lower than chloride threshold level, and after nitriteconcentration increased to 0.2 mol/L, the sulfate thresholdlevel became higher than chloride threshold level.This meansthat sulfate-induced corrosion was more likely to occur than chloride-induced corrosion in carbonated and neutral environments in which nitrite concentration was equal to or less than 0.053 mol/L. Whether sulfate threshold level was higher or lower than chloride threshold level in carbonated and neutral environments was determined by nitrite content.

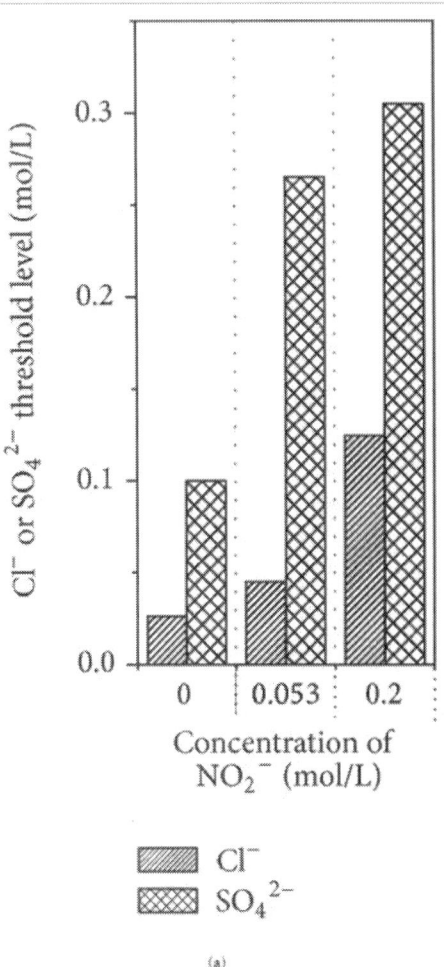

(a)

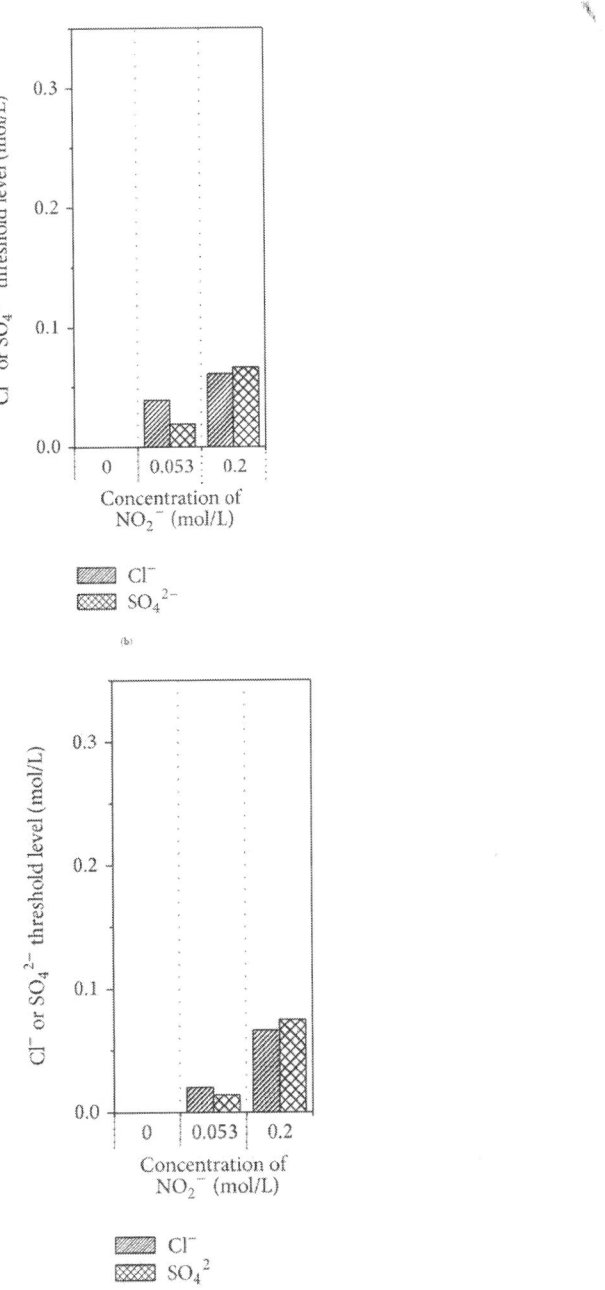

Figure 13: Comparing the threshold level of Cl– and SO$_4^{2-}$. (a) pH = 12.6, (b) pH = 10.3, and (c) pH = 8.1.

Effect on Anodic/Cathodic Polarization Curves, Tafel Slope, and Stern-Geary Constant

Anodic/cathodic polarizationcurves of steel were obtained based on the method described above. After the cathodic (anodic) polarization, some changes might occur on the surface of steel and thuswould have an influence on the further polarization of anodic(cathodic). In order to clarify this influence and give ananswer to this problem, a preliminary experiment was carriedout on the steel that had been immersed in distilled waterfor 3 days. Firstly, the cathodic polarization was done, and after finishing it, the steel was continued to be immersed in the solution about one day for the recovery of E_{corr}. And then, the anodic polarization was performed. The result was shown in Figure 14 and marked as black square "*". Afterthe cathodic and anodic polarization curves were obtained, the steel was continued to be immersed in the solution for the recovery of E_{corr}. About three days later, the cathodic and anodic polarization were performed again, and this time, the anodic polarization was done firstly, and one day later, the cathodic polarization curve was obtained. The result was also presented in Figure 14 and marked as red star "*". Comparing the results, little difference can be observed from the two polarization curves. Some changes of the surface state of steel might occur after the cathodic (anodic) polarization, but afterthe recovery of E_{corr}, they were not enough to affect the further polarization of anodic (cathodic).

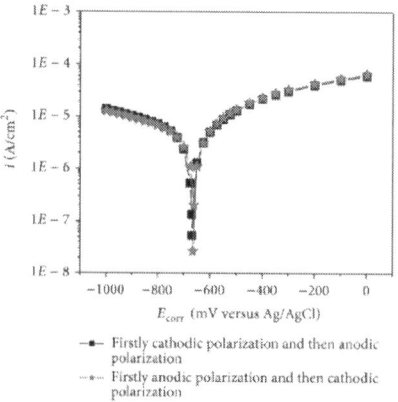

Figure 14: Polarization curves of steel immersed in distilled water for 3 days.

The effect of nitrite concentrations on the anodic and cathodic olarization curves, Tafel slope, and Stern-Geary constant B in pH 12.6, pH 10.3, and pH 8.1 environments with and without 0.1 mol/L Cl^- or SO_4^{2-} was shown in Figures 15and 16, respectively.

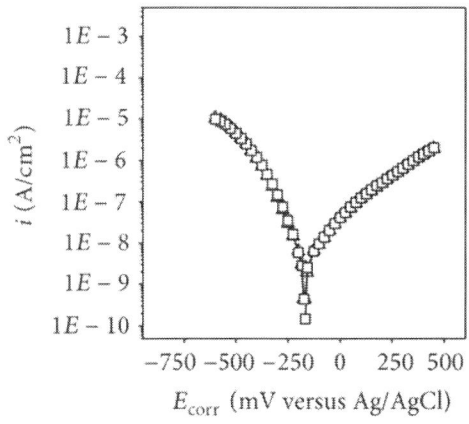

(a)

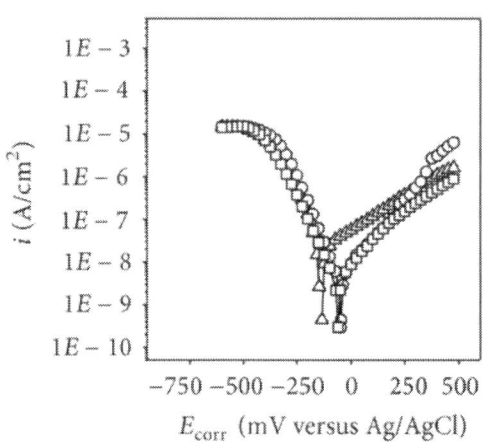

(b)

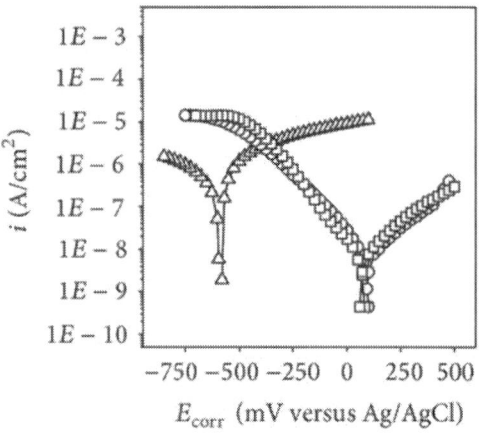

(c)

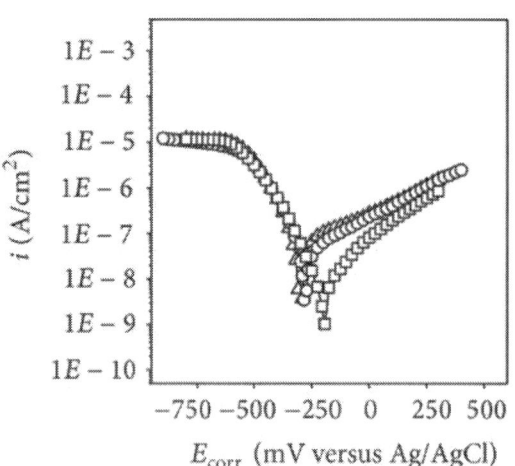

(d)

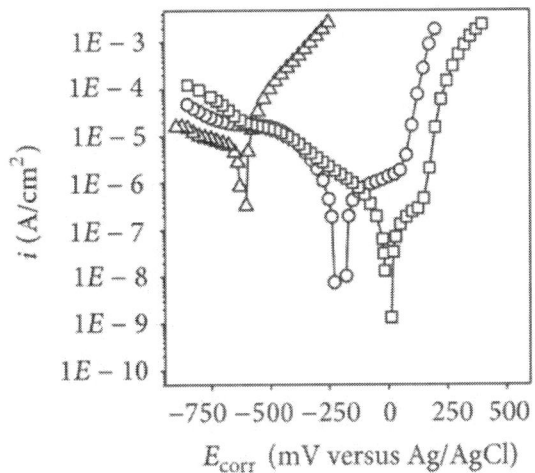

(f)

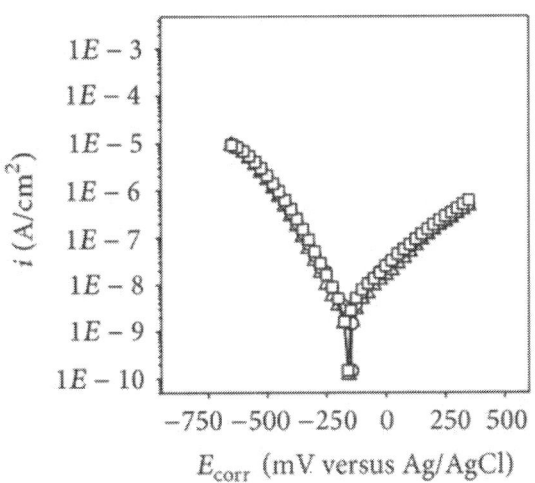

—△— $NO_2^- = 0$ mol/L
—○— $NO_2^- = 0.02$ mol/L
—□— $NO_2^- = 0.2$ mol/L

(g)

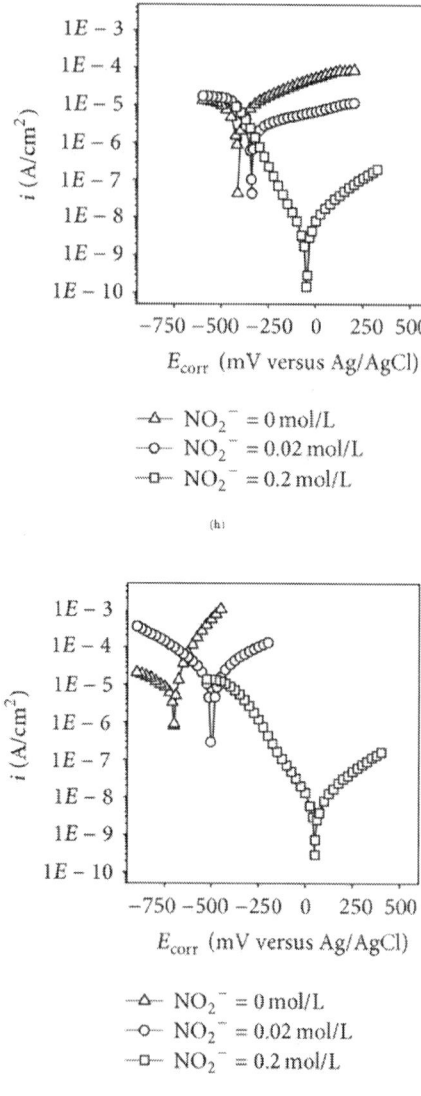

Figure 15: Effect of NO_2^- concentration on anodic and cathodic polarization curves of steel immersed in different pH solutions with and without 0.1 mol/L Cl⁻ or SO_4^{2-}. (a) pH 12.6 without Cl⁻ or SO_4^{2-}, (b) pH 10.3 without Cl⁻ or SO_4^{2-}, (c) pH 8.1 without Cl⁻ or SO_4^{2-}, (d) pH 12.6 with 0.1 mol/L Cl⁻, (e) pH 10.3 with 0.1 mol/L Cl⁻, (f) pH 8.1 with 0.1 mol/L Cl–, (g) pH 12.6 with 0.1 mol/L SO_4^{2-}, (h) pH 10.3 with 0.1 mol/LSO42–, and (i) pH 8.1 with 0.1 mol/L SO_4^{2-}.

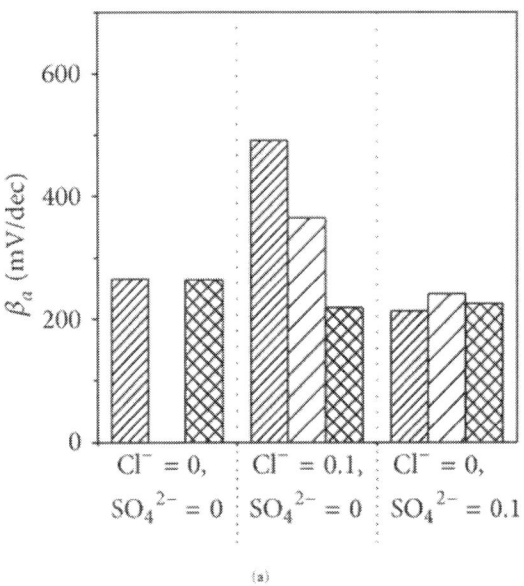

(a)

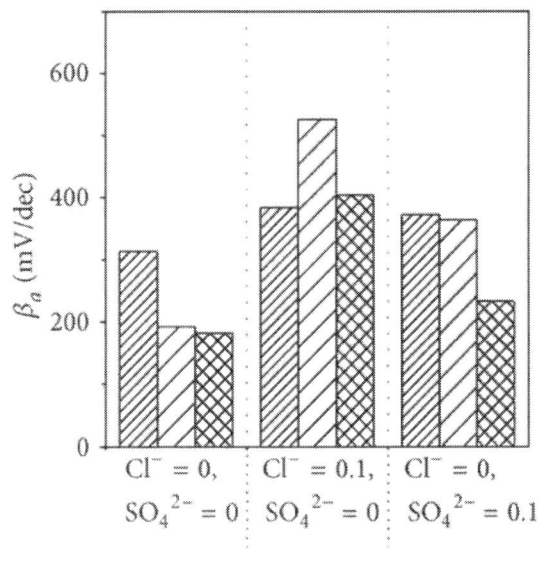

(b)

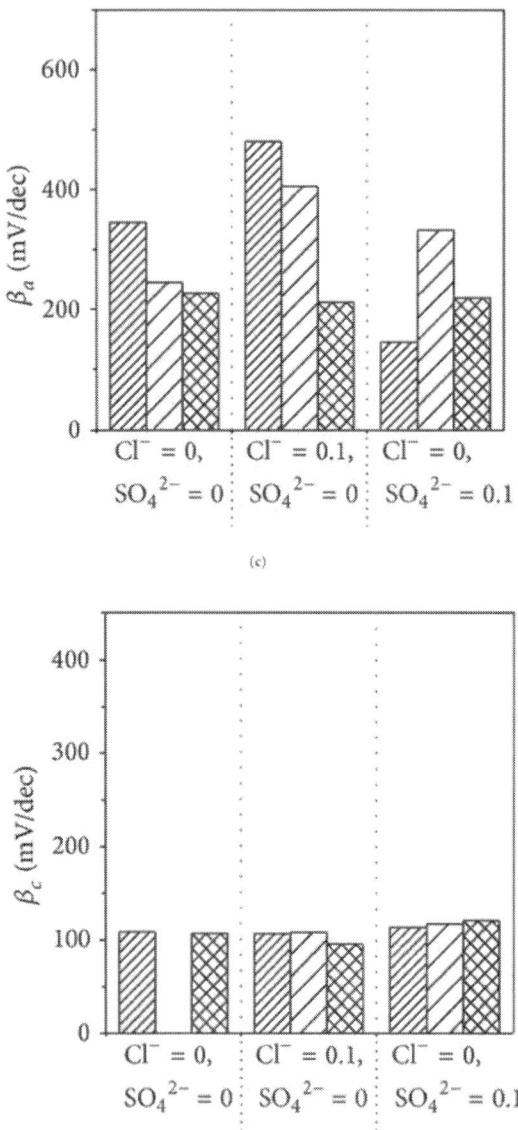

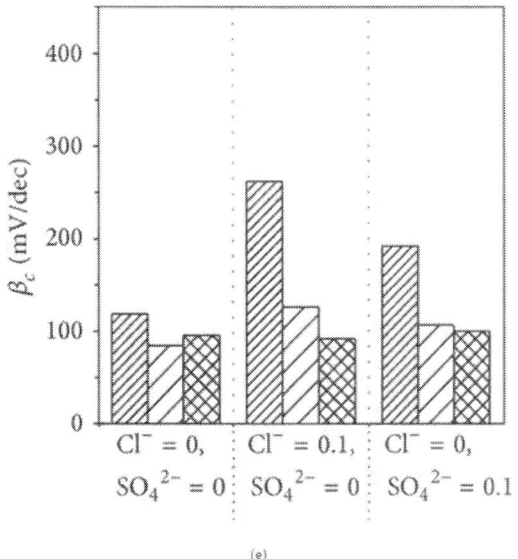

(e)

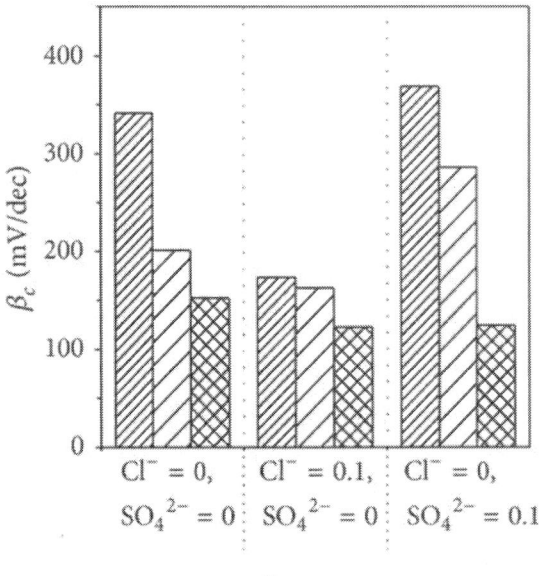

(f)

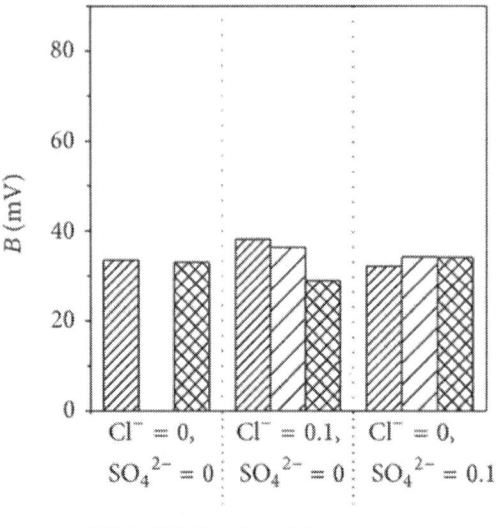

(g)

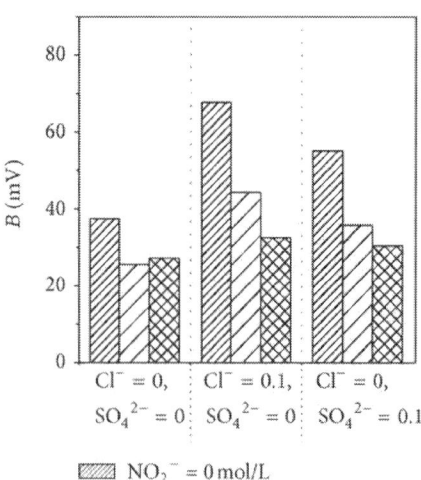

(h)

Guide to Stability Design Criteria for Metal Structures | 143

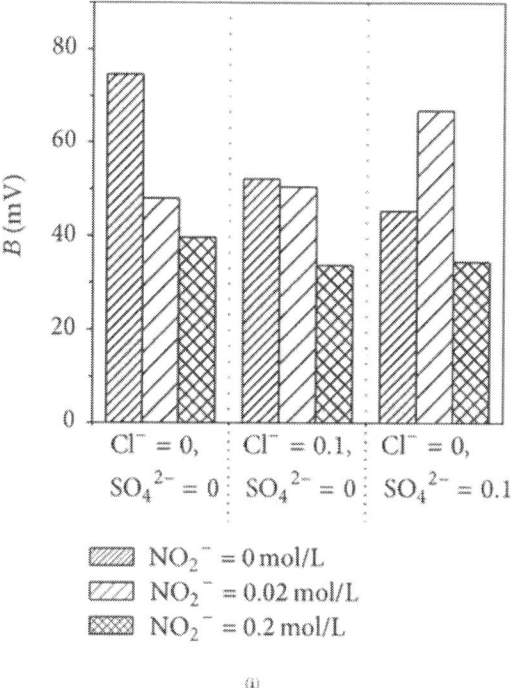

(a)

Figure 16: Effect of NO$_2^-$ concentration on Tafel slope and Stern-Geary constant B. (a) β_a in pH 12.6, (b) β_a in pH 10.3, (c) β_a in pH 8.1, (d) β_c in pH 12.6, (e) β_c in pH 10.3, (f) β_c in pH 8.1, (g) B in pH 12.6, (h) B in pH 10.3, and (i) B in pH 8.1.

Steels immersed in pH 12.6 environments without any chloride or sulfate (Figure 15(a)), and in the same solutions with 0.1 mol/L SO$_4^{2-}$ (Figure 15(g)) they were all in the passive state. With the presence of 0.02 and 0.2 mol/L nitrite, both the E_{corr} and polarization curves of steel did not show any obvious change. So the presence of nitrite had little influence on the anodic and cathodic polarization behaviors of steel. There was almost no change in β_a, β_c, and B in these conditions (Figures 16(a), 16(d), and 16(g)). However, in pH 12.6 environments with 0.1 mol/L Cl$^-$ (Figure 15(d)), the steel was corroded with the presence of 0 and 0.02 mol/L nitrite and was in passive state when nitrite concentration was 0.2 mol/L. With the increase of nitrite concentration, there was almost no change in the cathodic polarization curve, but the anodic polarization curves moved toward the direction of lower current density.

Corresponding to this, β_a decreased, while β_c remained constant, which resulted in the decrease of B value as described in Figures 16(a), 16(d), and 16(g).

In pH 10.3 environment without any Cl⁻ or SO_4^{2-} (Figure 15(b)), the steel was corroded with the absence of nitrite and was in passive state when nitrite concentration was 0.02 and 0.2 mol/L. With the increase of nitrite concentration, little difference in the cathodic polarization curves was observed. The presence of 0.02 mol/L nitrite made the anodic polarization curves shift towards the direction of lower current density, but higher nitrite concentration (0.2 mol/L) seemed to have no significant effect on the further decrease of current density of anodic polarization. Tafel slopes β_a and β_c and Stern-Geary constant B all decreased with the presence of nitrite as presented in Figures 16(b), 16(e), and 16(h)

In pH 10.3 environments with 0.1 mol/L Cl– (Figure 15(e)), the steel was corroded seriously with the absence of nitrite. Upon increasing the potential above the self-potential, a passive region was found where the current density was of the order of 10–5 A/cm². At potential value higher than −125 mV versus Ag/AgCl, an abrupt increase in the current density was observed resulting from the onset of the oxygen evolution reaction. The anodic polarization curves obtained in the presence of 0.02 mol/L nitrite were different from the one obtained in the absence of nitrite ions. The presence of higher nitrite contents had a substantial effect on the anodic behavior of the steel. The anodic polarization curves showed a passive zone, where the passive current density decreased with the increasing of nitrite concentrations. It was in the order of 10–6 A/cm2 for 0.02 mol/L nitrite and 10–7 A/cm2 for 0.2 mol/L nitrite. On the other hand, the passivity breakdown potential was observed and increased with the increasing of nitrite concentration. It was 125 mV for 0.02 mol/L nitrite and 225 mV for 0.2 mol/L nitrite. Besides this, Tafel slopes β_a and β_c and Stern-Geary constant B were greatly influenced by the nitrite concentration. The value of β_c and B decreased with the increasing of nitrite concentration. The B value with 0 mol/L nitrite was almost twice that with 0.2 mol/L nitrite as shown in Figures 16(b), 16(e), and 16(h).

In pH 10.3 environment with 0.1 mol/L SO_4^{2-} (Figure 15(h)), with the absence of nitrite, the steel was corroded. As the potential was raised from the self-potential, the current density increased with the potential and attained a value of the order of 10–4 A/cm2. No breakdown potential was observed even when the potential reached 200 mV versus Ag/AgCl. The presence of 0.02 and 0.2 mol/L nitrite ions not only shifted the self-potential toward higher values but also decreased the current density of anodic and cathodic polarization in the condition of the same over potential. The higher the nitrite concentration, the lower the current density of anodic and cathodic polarization. In this case, both Tafel slopes β_a and β_c and Stern-Geary constant B decreased with the presence of nitrite as given in Figures 16(b), 16(e), and 16(h). (b), 16(e), and 16(h).

In pH 8.1 environment without any Cl^- or SO_4^{2-} (Figure 15(c)), the steel was corroded seriously with the absence of nitrite, and it was in passive state when nitrite concentration was 0.02 and 0.2 mol/L. The presence of low content of nitrite (0.02 mol/L) not only greatly raised the steel potential and reduced the current density of anodic polarization but also markedly decreased the current densityof cathodic polarization. Under the same over potential, the cathodic polarizationcurrent density obtained in the solution containing 0.02 mol/L nitrite was about one order of magnitude lower than that obtained in the absence of nitrite, and further increase of nitrite concentration (0.2 mol/L) did not result in the further decrease of current density of cathodic polarization. In this case, Tafel slopes β_a and β_c and Stern-Geary constant B decreased with the presence of nitrite as presented in Figures 16(c), 16(f), and 16(i).

In pH 8.1 environments with 0.1 mol/L Cl^- (Figure 15(f)), the steel was corroded seriously with the absence of nitrite.Upon increasing the potential above the self-potential, nopassive region can be found. With the presence of nitrite,the self-potential shifted toward the positive directionand was function of nitrite concentration. The anodic polarization behavior of steel in pH 8.1 environments containing 0.1 mol/L Cl^- and 0.02 or 0.2 mol/L NO_2^- was found to be similar to that observed in pH 10.3 environment (Figure 15(e)). A passive region followed by an abrupt increase of current density at the potential higher than 50 mV for 0.02 mol/L nitrite or 150 mV for 0.2 mol/L nitrite was observed. The presence of nitrite also had an effect on the cathodic behavior of steel, which caused the cathodic polarization curve to shift to the direction of lower current density. In this case, β_a firstly increased when nitrite content was very low (0.02 mol/L) and then decreased when nitrite content was high (0.2 mol/L), while β_c and Stern-Geary constant B decreased with the presence of nitrite as shown in Figures 16(c), 16(f), and 16(i).

In pH 8.1 environment only containing 0.1 mol/L SO_4^{2-} (Figure 15(i)), the steel was corroded seriously with the absence of nitrite. The presence of 0.02 mol/L nitrite led to the increase of cathodic polarization current density and the decrease of anodic polarization current density. Higher nitrite concentration (0.2 mol/L) resulted in further decrease of anodic polarization current density and also had an influence on cathodic polarization current density. It can be seen that under the same over potential, the cathodic polarizationcurrent in the case of 0.2 mol/L nitrite was lower than that in the case of 0 mol/L nitrite. After the anodic and cathodicpolarization were finished, visual examination of steel surface was performed. Many small pitting holes with the sizes of 0.2 mm to 1 mm were observed on the surface of steels immersed in pH 8.1 solutions containing 0.02 mol/L nitrite and 0.1 mol/L SO_4^{2-}, while no rust and no pitting hole were found on the surface of steels immersed in pH 8.1 solutions containing 0.2 mol/L nitrite and 0.1 mol/L $SO4^{2-}$. Cathodic Tafel slope βc decreased with the presence of nitrite, while anodic Tafel slope βa and Stern-Geary constant B firstly increased when nitrite content was very low (0.02 mol/L) andthen decreased when nitrite content was high (0.2 mol/L) as shown in Figures 16(c), 16(f), and 16(i).

When steel is polarized anodically, the steel potential set by potentiostat is greater than self-potential. The electrons produced per unit time by the Fe → Fe_2^{++} 2e− reaction exceed those consumed per unit time by the O_2 + $2H_2O$ + 4e⁻ → 4OH− reaction, and net oxidation occurs at the steel surface. A positive value is consistent with the sign convention that assigns a positive value to the external circuit current. When steel is corroded, the presence of nitrite accelerates the oxidation of Fe_2^+ into Fe_2O^3 or γFeOOH and causes the formation of a less porous and more compact passive film which inhibits the Fe → Fe_2^{++} 2e− reaction. The reduction of electrons produced per unit time by the Fe → Fe_2^{++} 2e⁻ reaction results in the decrease of anodic polarization current. When steel is in passive state, the properties and composition of the passive film are similar to those found in the presence of nitrite [13], so the Fe → Fe_2^{++} 2e− reaction is little affected and the electrons assigned to the external circuit is not changed. Therefore, there is almost no change in anodic polarization current.

CONCLUSIONS

In this study, the effect of nitrite concentration and pH on the corrosion of steel has been investigated by means of visual examination, half-cell potential, and anodic and cathodic polarization curves. Based on the above analysis, the following conclusions can be obtained.

With the presence of nitrite, the corrosion of steel can be inhibited effectively. Chloride threshold level and sulfate threshold level are not only increased with the increasingof nitrite concentration but are also affected by pH. Highlyalkaline environment plays an important role in assistingnitrite to inhibit corrosion.

Chloride-induced corrosion is more prone to initiate than sulfate-induced corrosion in highly alkaline environment, but in neutral environment, when nitrite concentration is equal to or less than 0.053 mol/L, sulfate-induced corrosion is more likely to occur than chloride-induced corrosion.

For chloride-induced corrosion, using NO_2^-/Cl^- mole ratio as the parameter to guarantee the inhibition effect is suitable in highly alkaline environment, but not appropriate in weakly alkaline and neutral environments. For sulfate induced corrosion, NO_2^-/SO_4^{2-} cannot be used as the parameter to guarantee the inhibition effect due to the change of its threshold level with nitrite concentration.

When the steel is in passive state, the presence of nitrite has little influence on anodic/cathodic polarization curves and Stern-Geary constant B. When the steel is corroded, the addition of nitrite has a significant effect on the anodic polarization curve of steel and the Stern-Geary constant B.

REFERENCES

1. K. K. Sideris and A. E. Savva, "Durability of mixtures containing calcium nitrite based corrosion inhibitor," Cement and Concrete Composites, vol. 27, no. 2, pp. 277–287, 2005.
2. K. Y. Ann and N. R. Buenfeld, "The effect of calcium nitrite on the chloride-induced corrosion of steel in concrete," Magazine of Concrete Research, vol. 59, no. 9, pp. 689–697, 2007.
3. M. Balonis and F. P. Glasser, "Calcium nitrite corrosion inhibitor in portland cement: influence of nitrite on chloride binding and mineralogy," Journal of the American Ceramic Society, vol. 94, no. 7, pp. 2230–2241, 2011.

4. C. Alonso and C. Andrade, "Effect of nitrite as a corrosion inhibitor in contaminated and chloride-free carbonated mortars," ACI Materials Journal, vol. 87, no. 2, pp. 130–137, 1990.
5. V. T. Ngala, C. L. Page, and M. M. Page, "Corrosion inhibitor systems for remedial treatment of reinforced concrete.Part 1: calcium nitrite," Corrosion Science, vol. 44, no. 9, pp. 2073–2087, 2002.
6. M. B. Valcarce, C. López, and M. Vázquez, "The role of chloride, nitrite and carbonate ions on carbon steel passivity studied in simulating concrete pore solutions," Journal of the Electrochemical Society, vol. 159, no. 5, pp. C244–C251, 2012.
7. P. Montes, T. W. Bremner, and D. H. Lister, "Influence of calcium nitrite inhibitor and crack width on corrosion of steel in high performance concrete subjected to a simulated marine environment," Cement and Concrete Composites, vol. 26, no. 3, pp. 243–253, 2004.
8. L. Kondratova, P. Montes, and T. W. Bremner, "Natural marine exposure results for reinforced concrete slabs with corrosion inhibitors," Cement and Concrete Composites, vol. 25, no. 4-5, pp. 483–490, 2003.
9. M. B. Valcarce and M. Vázquez, "Carbon steel passivity examined in alkaline solutions: the effect of chloride and nitrite ions," ElectrochimicaActa, vol. 53, no. 15, pp. 5007–5015, 2008.
10. L. Mammoliti, C. M. Hansson, and B. B. Hope, "Corrosion inhibitors in concrete. Part II: effect on chloride threshold values for corrosion of steel in synthetic pore solutions," Cement and Concrete Research, vol. 29, no. 10, pp. 1583–1589, 1999.
11. Królikowski and J. Kuziak, "Impedance study on calcium nitrite as a penetrating corrosion inhibitor for steel in concrete," ElectrochimicaActa, vol. 56, no. 23, pp. 7845–7853, 2011.
12. P. Garcés, P. Saura, E. Zornoza, and C. Andrade, "Influence of pH on the nitrite corrosion inhibition of reinforcing steel in simulated concrete pore solution," Corrosion Science, vol. 53, no. 12, pp. 3991–4000, 2011.
13. M. B. Valcarce and M. Vázquez, "Carbon steel passivity examined in solutions with a low degree of carbonation: the effect of chloride and nitrite ions," Materials Chemistry and Physics, vol. 115, no. 1, pp. 313–321, 2009.
14. M. Reffass, R. Sabot, M. Jeannin, C. Berziou, and P. Refait, "Effects of NO_2^- ions on localised corrosion of steel in $NaHCO_3$ + $NaCl$ electrolytes," ElectrochimicaActa, vol. 52, no. 27, pp. 7599–7606, 2007.
15. P. Garcés, P. Saura, A. Méndez, E. Zornoza, and C. Andrade, "Effect of nitrite in corrosion of reinforcing steel in neutral and acid solutions simulating the electrolytic environments of micropores of concrete in the propagation period," Corrosion Science, vol. 50, no. 2, pp. 498–509, 2008.
16. S. Matsuda and H. H. Uhlig, "Effect of pH, sulfates, and chlorides on behavior of sodium chromate and nitrite as passivators for steel," Journal of Electrochemical Society, vol. 111, no. 2, pp. 156–161, 1964.

17. T. Cheng, J. Lee, and W. Tsai, "Corrosion of reinforcements in artificial sea water and concentrated sulfate solution," Cement and Concrete Research, vol. 20, no. 2, pp. 243–252, 1990.
18. J. Al-Tayyib and M. S. Khan, "Effect of sulfate ions on the corrosion of rebars embedded in concrete," Cement and Concrete Composites, vol. 13, no. 2, pp. 123–127, 1991.
19. O. S. B. Al-Amoudi and M. Maslehuddin, "The effect of chloride and sulfate ions on reinforcement corrosion," Cement and Concrete Research, vol. 23, no. 1, pp. 139–146, 1993.
20. N. R. Jarrah, O. S. B. Al-Amoudi, M. Maslehuddin, O. A. Ashiru, and A. I. Al-Mana, "Electrochemical behaviour of steel in plain and blended cement concretes in sulphate and/or chloride environments," Construction and Building Materials, vol. 9, no. 2, pp. 97–103, 1995.
21. H. A. F. Dehwah, M. Maslehuddin, and S. A. Austin, "Long-term effect of sulfate ions and associated cation type on chloride-induced reinforcement corrosion in Portland cement concretes," Cement and Concrete Composites, vol. 24, no. 1, pp. 17–25, 2002.
22. M. G. Pujar, T. Anita, H. Shaikh, R. K. Dayal, and H. S. Khatak, "Use of electrochemical noise (EN) technique to study the effect of sulfate and chloride ions on passivation and pitting corrosion behavior of 316 stainless steel," Journal of Materials Engineering and Performance, vol. 16, no. 4, pp. 501–506, 2007.
23. E. E. Abd El Aal, S. Abd El Wanees, A. Diab, and S. M. Abd El Haleem, "Environmental factors affecting the corrosion behavior of reinforcing steel III. Measurement of pitting corrosion currents of steel in $Ca(OH)_2$ solutions under natural corrosion conditions," Corrosion Science, vol. 51, no. 8, pp. 1611–1618, 2009.
24. M. Stern and A. L. Geary, "Electrochemical polarization. I. A theoretical analysis of the shape of polarization Curves," Journal of the Electrochemical Society, vol. 104, no. 1, pp. 56–63, 1957.
25. P. Ghods, O. B. Isgor, G. McRae, and T. Miller, "The effect of concrete pore solution composition on the quality of passive oxide films on black steel reinforcement," Cement and Concrete Composites, vol. 31, no. 1, pp. 2–11, 2009.
26. M. Moreno, W. Morris, M. G. Alvarez, and G. S. Duffó, "Corrosion of reinforcing steel in simulated concrete pore solutions effect of carbonation and chloride content," Corrosion Science, vol. 46, no. 11, pp. 2681–2699, 2004.
27. B. Huet, V. L'Hostis, F. Miserque, and H. Idrissi, "Electrochemical behavior of mild steel in concrete: influence of pH and carbonate content of concrete pore solution," ElectrochimicaActa, vol. 51, no. 1, pp. 172–180, 2005.
28. E. E. Stansbury, Fundamentals of Electrochemical Corrosion, pp. 233–254, ASM International, Materials Park, Ohio, USA, 2000.

29. M. J. Pryor, "The significance of the flade potential," Journal of the Electrochemical Society, vol. 106, no. 7, pp. 557–562, 1959.
30. H. J. Engell, "Stability and breakdown phenomena of passivating films," ElectrochimicaActa, vol. 22, no. 9, pp. 987–993, 1977.

CITATION

Zhonglu Cao, Makoto Hibino, and Hiroki Goda, "Effect of Nitrite Ions on Steel Corrosion Induced by Chloride or Sulfate Ions," International Journal of Corrosion, vol. 2013, Article ID 853730, 16 pages, 2013. doi:10.1155/2013/853730

Durability and Corrosion of Aluminium and Its Alloys: Overview, Property Space, Techniques and Developments

N. L. Sukiman[1,6], X. Zhou[1], N. Birbilis[1], A.E. Hughes[2], J. M. C. Mol[3], S. J. Garcia[4], X. Zhou[5] and G. E. Thompson[5]

[1] Department of Materials Engineering, Monash University, Clayton, Australia
[2] CSIRO Materials Science and Technology, Melbourne, Australia
[3] TU Delft, Materials Department, Delft, Netherlands
[4] TU Delft, Aerospace Engineering, Delft, Netherlands
[5] School of Materials, The University of Manchester, Manchester, United Kingdom
[6] Department of Mechanical Engineering, University of Malaya, Kuala Lumpur, Malaysia

INTRODUCTION

Aluminium (Al) is an important structural engineering material, its usage ranking only behind ferrous alloys (Birbilis, Muster et al. 2011). The growth in usage and production of Al continues to increase (Davis 1999). The extensive use of Al lies in its strength:density ratio, toughness, and to some degree, its corrosion resistance. From a corrosion perspective, which is most relevant to this chapter, Al has been a successful metal used in a number of applications from commodity roles, to structural components of aircraft. A number of Al alloys can be satisfactorily deployed in environmental/atmospheric conditions in their conventional form, leaving the corrosion protection industry to focus on market needs in more demanding applications (such as those which require coating systems, for example, the aerospace industry).

Relatively pure aluminium presents good corrosion resistance due to the formation of a barrier oxide film that is bonded strongly to its surface (passive layer) and, that if damaged, re-forms immediately in most environments; i.e. re-passivation (Davis 1999). This protective oxide layer is especially stable in near-neutral solutions of most non-halide salts leading to excellent pitting resistance. Nevertheless, in open air solutions containing halide ions, with Cl$^-$ being the most common, aluminium is susceptible to pitting corrosion. This process occurs, because in the presence of oxygen, the metal is readily polarised to its pitting potential, and because chlorides contribute to the formation of soluble chlorinated aluminium (hydr)oxide which interferes with the formation of a stable oxide on the aluminium surface.

Aluminium and its alloys readily oxidises, including when Al is present in either in solid solution or in intermetallic (IM) particles. Industrial alloy surfaces however, tend to be as heterogeneous as their underlying microstructures. The surface of a wrought or cast alloy is likely to contain not only aluminium oxide alone, but may for example contain a fragment of a mixed Al-Mg oxide for alloys rich in Mg (Harvey, Hughes et al. 2008)). This is primarily because of the heat of segregation of Mg is high and it has a favorable free energy for oxide formation. If however an Al surface is mechanically undisturbed - then the surface oxide is relatively protective. Though, most real surfaces have some sort of mechanical finishing which results in the formation of a near surface deformed layer (NSDL) and shingling. Shingling occurs where the alloy matrix is spread across the surface including IM particles in abrasion and milling (Scholes, Furman et al. 2006; Muster, Hughes et al. 2009). This is because the IM particles are harder than the surrounding matrix and less susceptible to deformation (Zhou, Liu et al. 2011). Even on polished surfaces, the matrix and the IM particles rapidly form different oxide structures (Juffs, Hughes et al. 2001; Juffs, Hughes et al. 2002). This is almost certainly due to different chemical environments and different electrochemical reactions over the IM particles compared to the matrix. Furthermore, the morphology and the oxide are not continuous from the IM particles to the matrix and this represents a potential defect site in the context of corrosion. For the purposes of descriptions herein, IM particles can be classified into three main types; i) precipitates, ii) constituent particles and iii) dispersoids. Precipitates are typically in the shape of needles, laths, plates or spherical with the size ranging from Angstroms to fractions of a micrometer. They are formed by nucleation and growth from a supersaturated solid solution during low temperature aging and may be concentrated along the grain boundaries. Constituent particles however, are relatively large with irregular shape and the size can be up to 10 micrometers. This type of particle forms during solidification of the alloy and is not fully dissolved by subsequent thermomechanical processing (including solution heat treatments). They can be found in colonies of several IM crystals or different compound types. On the contrary, dispersoids are small particles with size ranging from 0.05 to 0.5 micrometers. They are thermally stable intermetallics of a fine size that are functional for controlling grain size and recrystallisation behavior. Dispersoids form by low level additions of highly insoluble elements such as Cr, Mn or Zr.

This chapter will aim to cover some of the important aspects related to the corrosion of Al-alloys, bearing in mind the role of alloy chemistry. In addition, some of the topical aspects related to techniques and ongoing developments in the general field of Al-alloy corrosion are presented. An attempt has been made to give the reader an overview of the key technical aspects, but unfortunately for comprehensive insight into the topic overall, the size of this chapter alone cannot be a replacement to dedicated monographs on the specific topics at hand; nor the ever-evolving journal literature that represents the state of the art. To aid in the transfer of information, this chapter has been divided into a number of sections to treat the widely varying topics independently.

The General Performance of the Al-Alloy Classes

The corrosion potential of an aluminium alloy in a given environment is primarily determined by the composition of the aluminium rich solid solution, which constitutes the predominant volume fraction and area fraction of the alloy microstructure (Davis 1999). While the potential is not affected significantly by second phase IM particles of microscopic size, these particles frequently have corrosion potentials (when measured in isolation) differing from that of the solid solution matrix resulting in local (micro-) galvanic cells, when IMs are polarised to the corrosion potential of the alloy. The result is that local currents on the alloy surface differ, establishing anodes and cathodes. Since most of the commercial aluminium alloys contain additions of more than one type of alloying element, the effects of multiple elements on solution potential are approximately additive. The amounts retained in solid solution, particularly for more highly alloyed compositions, depend on production and thermal processing so that the heat treatment and other processing variables influence the final electrode potential of the product.

By measuring the potentials of grain boundaries and grain bodies separately, the difference in potential responsible for local types of corrosion such as intergranular corrosion, exfoliation, and stress corrosion cracking (SCC) can be quantified (Guillaumin and Mankowski 1998; Zhang and Frankel 2003). By measuring the corrosion potential of IMs (Buchheit 1995), and indeed by measurement of the polarisation response of IMs, even more significant insights into localised corrosion can be gained (Birbilis and Buchheit 2005). Such specialist topics are not dealt with in their entirety herein, however an abridged written summary of the performance of the key Al-alloy classes (as outlined by the Aluminium Association (Hatch 1984)) is provided below.

1XXX Series Alloys
Corrosion resistance of aluminium increases with increasing metal purity, however the use of the >99.8% grades is usually confined to those applications where very high corrosion resistance or ductility is required. In regards to such specialist applications however, the actual number of applications are very few. Consequently 1xxx series alloys are not commonly used or sold (but do serve as important feedstock to secondary alloy producers or production). In the instance where general-purpose alloys for lightly stressed applications are required, such alloys are approximately 99% pure aluminium and offer adequate corrosion resistance in near neutral environments. 1xxx is also sometimes used in cladding for example AA1230 is used as clad on AA2024 (Hatch 1984)

2XXX Series Alloys

Copper is one of the most common alloying additions - since it has appreciable solubility and can impart significant strengthening by promotion of age hardening (in fact, the Al-Cu system was the classical/original age hardening system (Hatch 1984)). These alloys were the foundation of the modern aerospace construction industry and, for example AA2024 (Al-4.4Cu-1.5Mg-0.8Mn), which is still used in many applications to this day, can achieve strengths in excess of 500MPa depending on temper (Polmear 2006).

3XXX Series Alloys

The 3xxx series alloys are a commodity product that is nominally available in the form of thin sheet (for beverage can usage). The key alloying element, Manganese, has a relatively low solubility in aluminium but can improve corrosion resistance when remaining in solid solution. Additions of manganese of up to about 1% form the basis of the non-heat treatable wrought alloys with good corrosion resistance, moderate strength (i.e. AA3003 tensile strength ~110MPa) and exceptionally high formability (Polmear 2006).

5XXX Series Alloys

Magnesium has significant solubility in aluminium and imparts substantial solid solution strengthening (which can also contribute to enhanced work hardening characteristics) (Davis 1999; Polmear 2006). The 5xxx series alloys (containing <~6% Mg) do not age harden. Whilst it is possible for β-phase (Mg_2Al_3) to precipitate in systems with above ~3%Mg, the β-phase is not a strengthening precipitate and actually weakens the alloy (by depleting the solute of Mg). Nominally, the corrosion resistance of 5xxx series alloys is good and their mechanical properties make them ideally suited for structural use in aggressive conditions (such as marine vessels). Fully work-hardened AA5456 (Al-4.7Mg-0.7Mn-0.12Cr) has a tensile strength of >380MPa. One corrosion issue with fully work-hardened 5xxx series alloys is that the heavy dislocation density (and supersaturation of the solid solution with Mg) can permit the sensitization of the microstructure by precipitation of deleterious β-phase (Mg_2Al_3) during sustained high temperature exposure (i.e. in service) (Baer, Windisch et al. 2000; Searles, Gouma et al. 2002; Davenport, Yuan et al. 2006; Goswami, Spanos et al. 2010).

6XXX Series Alloys
Silicon additions alone can lower the melting point of aluminium whilst simultaneously increasing fluidity (which is why the vast majority of cast Al products contain various amounts of Si). These alloys are increasing in importance in automotive applications for engine and drive train components – however are yet to realise the majority of market share. Heat-treatable Al-Mg-Si are predominantly structural materials (strengths >300MPa are possible), all of which have an appreciable resistance to corrosion, immunity to SCC and are weldable. To date, 6xxx series alloys are mainly used in extruded form, although increasing amounts of sheet are being produced (Birbilis, Muster et al. 2011). Magnesium and silicon additions are made in balanced amounts to form quasi-binary Al-Mg$_2$Si alloys, or excess silicon additions are made beyond the level required to form Mg$_2$Si. Alloys containing magnesium and silicon in excess of 1.4% develop higher strength upon aging.

7XXX Series Alloys
The Al-Zn-Mg alloy system provides a range of commercial compositions, primarily where strength is the key requirement (and this can be achieved without relatively high cost or complex alloying). Al-Zn-Mg-Cu alloys have traditionally offered the greatest potential for age hardening and as early as 1917 a tensile strength of 580MPa was achieved, however, such alloys were not suitable for commercial use until their high susceptibility to stress corrosion cracking could be moderated (Song, Dietzel et al. 2004; Birbilis, Cavanaugh et al. 2006; Lin, Liao et al. 2006; Lynch, Knight et al. 2009). Aerospace needs led to the introduction of a range of high strength aerospace alloys of which AA7075 (Al-5.6Zn-2.5Mg-1.6Cu-0.4Si-0.5Fe-0.3Mn-0.2Cr-0.2Ti) is perhaps the most well-known, and which is now essentially wholly superseded by AA7150 (or the 7x50 family). The high strength 7xxx series alloys derive their strength from the precipitation of η-phase (MgZn$_2$) and its precursor forms. The heat treatment of the 7xxx series alloys is complex, involving a range of heat treatments that have been developed to balance strength and stress corrosion cracking performance - including secondary (or more) heat

treatments that can include retrogression and re-aging (Fleck, Calleros et al. 2000; Ferrer, Koul et al. 2003; Zieliński, Chrzanowski et al. 2004; Marlaud, Deschamps et al. 2010).

8XXX Series Alloys
Nominally reserved for the sundry alloys, 8xxx series alloys include a number of Lithium (Li) containing alloys. Li is soluble in aluminium to ~ 4 wt% (corresponding to ~ 16 at%). As these alloys of high specific strength and stiffness readily respond to heat treatment, research and development has intensified due to their potential for widespread usage in aerospace applications (Lavernia and Grant 1987; Dorward and Pritchett 1988; Giummarra, Thomas et al. 2007). Based on the impressive lightweight of such alloys, present day aircraft are comprised of some portion of Al-Li based alloys (modern generations of which actually include low Li levels and hence are nowadays designated as 2xxx alloys (Ambat and Dwarakadasa 1992; Garrard 1994; Semenov 2001; Giummarra, Thomas et al. 2007). First generation Li-containing alloys displayed some of the highest corrosion rates of all aluminium alloys, where susceptibility to intergranular corrosion was challenging. Modern Al-Cu-Li seem to have overcome this challenge; however it is also important to recognise that production requires specialised melting and casting, not presently available in most commercial facilities.

CORROSION OF ALUMINIUM AND ITS ALLOYS IN AQUEOUS ENVIRONMENT

Environmental Corrosion of Aluminium

Corrosion in aluminium alloys is generally of a local nature, because of the separation of anodic and cathodic reactions and solution resistance limiting the galvanic cell size. The basic anodic reaction is metal dissolution:

$$Al \rightarrow Al^{3+} + 3e^-$$

While the cathodic reactions are oxygen reduction:

$$O_2 + 2H_2O + 4e^- \rightarrow 4OH^-$$

or hydrogen reduction in acidified solution such as in a pit environment as a result of aluminium ion hydrolysis:

$$2H^+ + 2e \rightarrow H_2$$

It is the interaction between local cathodes and anodes and the alloy matrix that leads to nearly all forms of corrosion in aluminium alloys. These include pitting corrosion, selective dissolution, trenching, intermetallic particle etchout, intergranular attack and exfoliation corrosion. Surface and subsurface grain etchout is also influenced by grain energy which is derived from grain defect density. Grain etchout, has a significant role in exfoliation corrosion since the volume of hydrated aluminium oxide generated during dissolution is larger than the original volume of the grain.

The general consensus for Al and its alloys is that they are resistant towards corrosion in mildly aggressive aqueous environments. The protective oxide layer represents the thermodynamic stability of Al alloys in corrosive environment - acting as a physical barrier as well as capable to repair itself in oxidizing environments if damaged. While the passive layer breakdown mechanism by chloride ions is still in debate (Sato 1990; McCafferty 2010) due to the complexity of the process (Szklarska-Smialowska 2002), the general consensus is that localized attack starts by adsorption of aggressive anions and formation of soluble transitional complexes with the cations at the oxide surface. Thermodynamic principles to explain and predict the passivity phenomenon that controls the corrosion behavior of Al are summarised by Pourbaix-type analysis. This results in a plot of potential vs. pH based on the electrochemical reaction of the species involved, the representation known as a Pourbaix diagram (Pourbaix 1974) as shown in Figure 1.

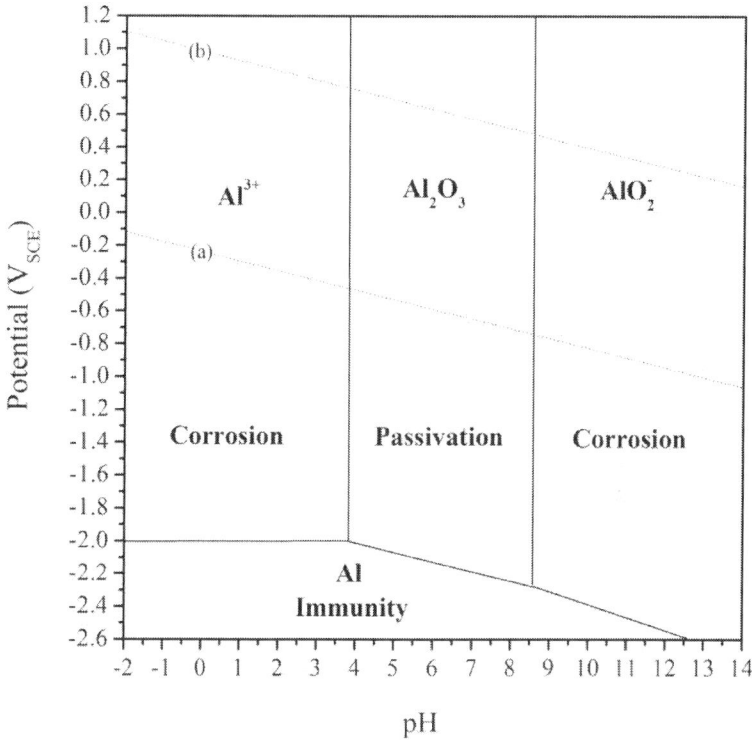

Figure 1: E-pH diagram for pure Al at 25°C in aqueous solution (adapted from Pourbaix 1974). The lines (a) and (b) correspond to water stability and its decomposed product.

It is seen that Al is nominally passive in the pH range of ~4 to 9 due to the presence of an Al_2O_3 film. In environments that deviate from the near neutral range, the continuity of this film can be disrupted in which the film becomes soluble, facilitating the relatively rapid of dissolution the alloy. In the acidic range, Al is oxidised by forming Al^{3+}, whilst AlO^{2-} occurs in alkaline range.

The E-pH diagram gives an impression that corrosion prediction is a straightforward process, however in actual engineering applications, there are several variables that weren't considered by Pourbaix. These include (i) the presence of alloying elements in most engineering metals (ii) the presence of substances in the electrolyte such as chloride (albeit that this has been addressed in more modern computations), (iii) the operating temperature of the alloy, (iv) the mode of corrosion, and (v) the rate of reaction. Taking these factors into account is nominally done on a case by case (i.e. alloy by alloy) basis, and a revised version of an E-pH diagram for 5xxx series alloys in 0.5M sodium chloride is given in Figure 2.

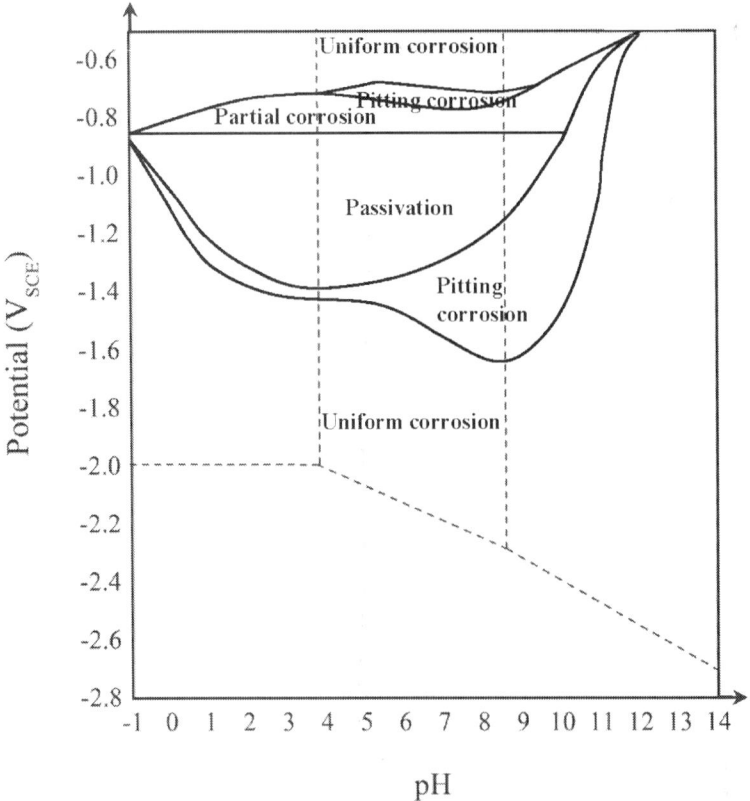

Figure 2: Mode of corrosion based on experimental data for AA5086 in the presence of 0.5M sodium chloride (adapted from Gimenez, Rameau et al. 1981)

Figure 2 indicates windows where localized attack is highly possible in the supposed passive region (Gimenez, Rameau et al. 1981). It is also seen that localised attack is possible across the whole range of pH depending on the specific potential. One should therefore not rely solely on the Pourbaix diagram as a direct index to actual corrosion rates, with rates needing to be independently measured for a given alloy-electrolyte combination (Ambat and Dwarakadasa 1992). Finally, whilst not to be discussed in detail here, it is prudent to indicate that effectively all Al-alloys do not attain practical/empirical immunity as evidence in Figure 1. Cathodic polarisation tends to contribute to alloy deterioration by two modes. Firstly, the accumulation of hydroxyl ions at the Al-surface will cause chemical dissolution of the Al. Secondly, Al is a very strong hydride former, and hydrogen from the cathodic reaction at such negative potentials will serve combine with Al to form hydrides (Perrault 1979).

Kinetic Stability of Aluminium Alloys

Kinetics represents the rate of reaction during corrosion. When exposed to an aqueous environment, metals stabilise to a value of electrochemical potential that is characteristic of the material and its composition for a given electrolyte. This potential is the potential at which anodic and cathodic reactions upon the metal surface are equal, and the value of this potential is thus significantly influenced by factors that can alter the relative rates of anodic or cathodic reaction efficiency upon the metal surface (i.e. alloying, precipitate state, etc.).

Most typically, the potentiodynamic polarisation test is used to characterise the corrosion performance of an alloy (as far as determination of mechanistic aspects from an instantaneous test). This method gives vital kinetic information such as current density over a range of potentials, pitting potential (if it exists), corrosion potential, the passive current density and potentially more information in reverse scans, etc. Thus factors affecting corrosion as discussed in the previous sections can be evaluated with much higher confidence. For example, Figure 3 shows a polarisation curve of Pure Al compare to AA2024-T3 (Al-4.3Cu-1.5Mg-0.6Mn) in 0.1M NaCl.

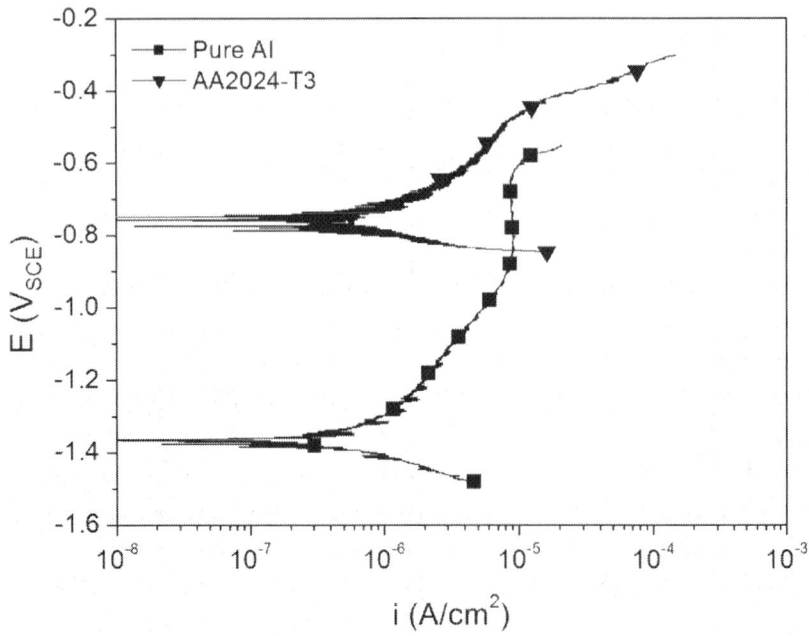

Figure 3: Polarisation curve of pure Al and AA2024-T3 exposed to 0.1M NaCl for 7 days collected at 1mV/s^{-1}(adapted from (Sukiman, Birbilis et al. 2010))

Comparison of alloy behavior and dominant reactions can be made in a quantitative manner. The anodic branch of the polarisation curve gives information related to the anodic/dissolution reaction while the cathodic branch represents the reduction reaction (nominally oxygen reduction, but at lower potentials or in strong acids, water reduction). The addition of more noble alloying elements typically increases the corrosion potential to more noble values (Davis 1999) and this is dramatically observed in Figure 3. This ennoblement however does not correlate to the rate of corrosion (as judged by Figure 3), whereby we see the pure Al versus AA2024-T3 has a difference in potential of ~0.5V. In addition, the main practical threat for Al alloys is localised attack such as pitting, so in that vein, a more noble value of pitting potential does not necessarily signify a better corrosion resistance (Frankel 1998; Birbilis and Buchheit 2005). Rather generally, the electrochemical reactions upon Al-alloys are heavily influenced by the chemistry and microstructure of the alloy – which we attempt to discuss in the following section.

Moving beyond potentiodynamic polarisation towards a true measure of kinetic stability in the E-pH domain (similar in concept to Pourbaix diagrams however which give 'speed' and not just thermodynamic likelihood) there are tests which can be done in this regard. In order to develop an improved understanding of overall kinetic stability of a metal over the potential-pH space, methods including the staircase potentio-electrochemical impedance spectroscopy (SPEIS) can be used to establish so-called kinetic stability diagrams, as previously demonstrated for pure Al (Zhou, Birbilis et al. 2010) and depicted in Figure 4. The specifics of SPEIS will be introduced in section 5.3.

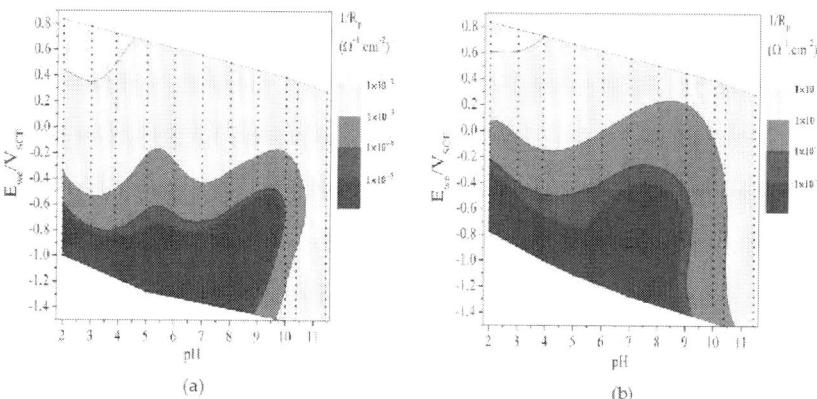

Figure 4: Contour plots of 1/RP for the E-pH space and their data for (a) A portion of a 99.9999% Al ingot single crystal and (b) a polycrystalline specimen from the identical ingot in (a) (adapted from (Zhou, Birbilis et al. 2010)

In Figure 4, presentation of the reciprocal of polarisation resistance (1/R_P) is the metric of reaction 'speed', as it is proportional to the reaction rate at a given condition. The influence of potential and pH is presented not only for pure Al in the sample which was single crystal, but also for a polycrystalline sample. As a result, one is also able to assess the effect of grain structure from Figure 4. Most importantly however, Figure 4 reveals that the rate of reaction stays in a low range in areas that extend beyond that of the Pourbaix diagram in terms of certain E-pH combinations – indicating that although Al may be in a thermodynamically unstable region, the reaction rate can be maintained to be low enough to make it still be useful in an engineering context. Similarly, there are regions of high potential where pure aluminium may be in a thermodynamically stable region, but unusable – owing to transpassive dissolution. Finally, in terms of microstructure effects, it is seen that the exact same material can have a different kinetic response based on structural factors alone. Such differences are not detectable or predicted from thermodynamic analysis in any way, and this highlights the importance of approaches which provide kinetic information to meet demands of engineering applications.

The Property Space and Corrosion Property Profile of Aluminium Alloys

As technologies continue to advance with more challenging applications and environments, a general understanding of durability limits across the class of Al-alloys is essential. Durability needed in a broad sense is the ability to withstand an environment while maintaining mechanical integrity. This indicates a requirement to understand the property space for Al and its alloys. Figure 5 shows a trend that is in line with the notion that increases in hardness (used here as a proxy to yield strength) show increases in corrosion rate.

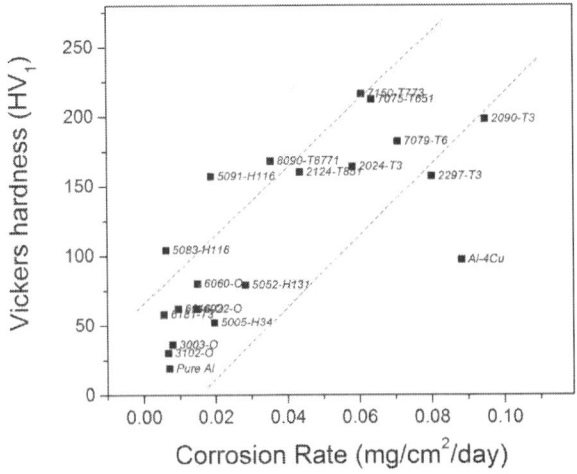

Figure 5: Corrosion rate as determined from weight loss data for commercial Al alloys collected after 14 days exposure in quiescent 0.1M NaCl presented against alloy hardness.

From Figure 5, it can be argued that the data is divided into two main groups, one at each end of the corrosion rate spectrum. High hardness/strength aluminium alloys were found to populate the high corrosion rate space. This is in contrast to the medium to low hardness/strength alloys that revealed considerably lower corrosion rates. The alloys that show the highest corrosion rates are the 'precipitation hardenable' family. Besides the high number density of precipitate particles in such alloys, they also contain an appreciable population of constituent type particles (Chen, Gao et al. 1996;Wei 2001; Andreatta, Terryn et al. 2004; Ilevbare, Schneider et al. 2004; Birbilis, Cavanaugh et al. 2006; Boag, Hughes et al. 2009; Hughes, MacRae et al. 2010; Xu, Birbilis et al. 2011). These particles are industrially necessary, since the complex chemistry of precipitation hardenable alloys (that can contain up to 10 alloying elements) have significant alloying additions added via alloy rich master alloys. It is also observed that the alloys that show the highest corrosion rates also contain appreciable amount of copper.

The plot in Figure 5 allows one to identify a region of property space (at the top left corner) that has potential for future alloys, with ongoing efforts aiming to reach such space (in addition to controlling ductility). Efforts that regard in are underway, focusing on corrosion rate (Carroll, Gouma et al. 2000;Norova, Ganiev et al. 2003; Rosalbino, Angelini et al. 2003; Cavanaugh, Birbilis et al. 2007; Lucente and Scully 2007; Fang, Chen et al. 2009; Graver, Pedersen et al. 2009; Ralston, Birbilis et al. 2010; Tan and Allen 2010; Xu, Birbilis et al. 2011; Brunner, Birbilis et al. 2012) and strength (Poole, Seter et al. 2000; Pedersen and Arnberg 2001; Fuller, Krause et al. 2002; Raviprasad, Hutchinson et al. 2003; Lee, Shin et al. 2004; Oliveira Jr, de Barros et al. 2004; Zhao, Liao et al. 2004; Kim, Kim et al. 2005;Teixeira, Bourgeois et al. 2007; McKenzie and Lapovok 2010; Wang, Zhang et al. 2010; Puga, Costa et al. 2011; Zhong, Feng et al. 2011; Westermann, Hopperstad et al. 2012). However, such studies are done independently of both properties thus the symbiotic effect can't be readily evaluated to date.

CORROSION OF ALUMINIUM AND ITS ALLOYS IN AQUEOUS ENVIRONMENT

The Role of Chemistry on Corrosion

Alloying elements are added to aluminium for various reasons, with improving mechanical properties the principal reason. These elements introduce heterogeneity into the microstructure, which is the main cause of localised corrosion that initiates in the form of pitting. Each alloying element has a different effect on the corrosion of Al, and in this section we *briefly* discuss the role of alloying elements on corrosion of Al.

Influence of Magnesium

Mg is one of the major elements added to Al to improve mechanical properties by solid solution strengthening – and can be found in 5xxx alloys, as well as 2xxx, 3xxx, 6xxx and 7xxx commercial alloys. Mg can stabilize GP zones, has a high solubility in Al and decreases the alloy density. Muller and Galvele showed that Mg when present in solid solution does not have a significant effect on the pitting corrosion of Al which can be understood on the basis of standard potentials of Al and Mg (Muller and Galvele 1977). Moreover, Mg decreases the rate of the cathodic reaction when present in solid solution, increasing corrosion resistance, which may appear counterintuitive, but is rather obvious (as Mg has a very low exchange current density and hence retards the cathodic reaction). In contrast, excess amounts of Mg in the alloy or a long term exposure to elevated temperature will cause the precipitation of either Al_8Mg_2 or Al_3Mg_2 (Searles, Gouma et al. 2002; Davenport, Yuan et al. 2006;Oguocha, Adigun et al. 2008; Jain, Lim et al. 2012). These phases form typically along grain boundaries (Baer, Windisch et al. 2000; Goswami, Spanos et al. 2010) and are known to be anodic with respect to Al matrix therefore prone to localized corrosion (Vetrano, Williford et al. 1997; Aballe, Bethencourt et al. 2001; Jones, Baer et al. 2001; Brunner, May et al. 2010). Mg in 2xxx, 6xxx and 7xxx alloys however, forms precipitates with other alloying elements to strengthen the alloy where role of Mg mainly depends upon the other alloying additions (Ringer, Hono et al. 1996; Buchheit, Grant et al.

1997; Campestrini, van Westing et al. 2000; Guillaumin and Mankowski 2000; Eckermann, Suter et al. 2008).

Influence of Silicon
The addition of Si in conjunction with Mg, which is typical in 6xxx series Al alloys, allows Mg$_2$Si particles to precipitate. There is vast literature on the chemical composition of the Mg-Si phase and its role on mechanical properties (Hirth, Marshall et al. 2001; Usta, Glicksman et al. 2004; Stelling, Irretier et al. 2006; Eckermann, Suter et al. 2008; Zeng, Wei et al. 2011). This particle is beneficial in terms of increasing strength but renders the alloy prone to localised corrosion (Eckermann, Suter et al. 2008). The electrochemical behavior of Mg$_2$Si was investigated recently and it was shown that Mg$_2$Si remains more 'anodic' (i.e.. less noble) than the matrix in Al-alloys. As a consequence of this, Mg$_2$Si remains anodic and undergoes selective dissolution in the Al-matrix. Some 6xxx series alloys contain excess Si. Excess amount of Si however increases the cathodic reaction rate (Eckermann, Suter et al. 2008) and are unfavorable since Si tends to be present along the grain boundary and this may lead to intergranular corrosion and stress corrosion cracking (Guillaumin and Mankowski 2000; Larsen, Walmsley et al. 2008; Zeng, Wei et al. 2011).

Influence of Copper
The presence of Cu is viewed as detrimental to corrosion due to the formation of cathodic particles capable of sustaining the cathodic reaction locally and efficiently, such as Al$_2$Cu and AlCu$_2$Mg. In some cases where low Cu content is used, the impact of Cu is minimal, however given that corrosion is not the principal alloy design criteria in most instance, Cu is common in many (most) Al-alloys. The 2xxx series alloys are Cu rich, however Cu is added to other alloy classes such as the 6xxx series where it can increase strength when present in trace amounts, and also enhance precipitation hardening. The same is true in 7xxx alloys, with most modern aerospace alloys having appreciable amounts of Cu that can increase strength by modifying precipitation and minimising SCC via incorporation into precipitates (such as Mg(Zn,Cu)$_2$).

In general however, there is still some debate on the precise role of Cu, which also depends on the temper condition. Muller and Buchheit found that Cu in the form of solid solution decreases pitting susceptibility through the ennoblement of pitting potential. While Muller and Galvele reported an increase in pitting potential for solid solution content of Cu up to 5 wt%. In the case of Al-Cu-Mg alloys which contain S phase (Al_2CuMg), large differences in solution potential between Cu (highly noble) and Mg exist, with significant focus on corrosion of S phase (Buchheit, Grant et al. 1997;Guillaumin and Mankowski 1998; Buchheit, Montes et al. 1999; Ilevbare, Schneider et al. 2004; Boag,Hughes et al. 2011) revealing dealloying and selective dissolution that leads to preferential dissolution of Mg and Al with Cu remnant being redistributed at or near the site of the Al_2CuMg. A range of other particles associated with Cu have been reported such as Al_7Cu_2Fe. However recent microprobe studies of a number of batches of AA2024-T351 indicate five common compositions across modern alloys which do not have the same composition as older alloy stock indicating that this is still an active area of research (Hughes, Glenn et al. 2012). In general, Cu, or Cu containing particles are capable of supporting high oxygen reductions rates and hence undesirable from corrosion perspective (Mazurkiewicz and Piotrowski 1983; Scully, Knight et al. 1993; Buchheit 2000; Birbilis, Cavanaugh et al. 2006).

Influence of Zinc

In high strength commercial aluminium alloys such 7xxx series alloys, Zn is added to stimulate precipitation hardening. Alloys containing high levels of Zn such as the modern aerospace alloys 7050 and 7150 are amongst the highest strengths of Al-alloys owing to the high number density of precipitates such as $MgZn_2$ which is evenly distributed throughout the Al matrix (Ringer, Hono et al. 1996; Andreatta, Terryn et al. 2004; Sha and Cerezo 2004; Birbilis and Buchheit 2005; Polmear 2006) in 5xxx alloys. The addition of Zn to Al-Mg alloys was reported to improve resistance against SCC (Unocic, Kobe et al. 2006) where a small amount of Zn added into AA5083 alloy was found to reduce the corrosion - reporting that Zn can promote the formation of Al-Mg-Zn (τ phase) instead of Al_3Mg_2(β phase) the latter of which is responsible for stress corrosion cracking (Carroll, Gouma et al. 2000; Carroll, Gouma et al. 2001).

Influence of Iron
Iron is typically present as an impurity in all commercial Al alloys due to the production process of Al alloys. Unless specifically required for specialist applications, it is simply too expensive to remove all iron (even in aluminium destined for aerospace applications). Despite having a small fraction of the composition, iron is detrimental to corrosion due to its low solubility and hence ability to form constituent particles which are cathodic to the Al-matrix such as Al_3Fe (Nisancioglu 1990). Additionally, iron is capable of sustaining cathodic reactions more efficiently than Al (Galvele 1976;Szklarska-Smialowska 1999). In more complex alloys, Fe can also combine with other alloying elements such as Mn or Cu (in the latter case forming Al_7Cu_2Fe), which are also a major issue for corrosion (Birbilis, Cavanaugh et al. 2006) since the combination of Fe and Cu provides even greater cathodic efficiency for such particles. Corrosion associated with such noble cathodic constituents/intermetallics leads to an increase in local pH of the solution further enhancing anodic dissolution of the Al matrix adjacent to say, Al_3Fe (Seri 1994; Park, Paik et al. 1999; Birbilis and Buchheit 2005; Ambat, Davenport et al. 2006).

Influence of Manganese
The addition of Mn is effective in reducing the pitting susceptibility of Al alloys particularly in the context of modifying Fe containing intermetallic particles (Nisancioglu 1990) (where Mn can substitute for Fe, rendering the resulting constituent particle somewhat less noble) The presence of Mn has been noted as reducing the concentration of Fe and reducing the degree of resultant corrosion (Koroleva, Thompson et al. 1999); owing to the formation of Al_6MnFe has a similar electrochemical potential with that of the Al matrix. However, it has also been noted that an excess of Mn can lead to an increase in the cathodic activity when beyond the solubility limit (solubility of Mn in Al is 1.25 wt%) – with constituents such as Al_6Mn readily forming (Liu and Cheng 2010). Generally however, the presence of Mn constituent particles are not as detrimental as particles rich in Fe or Cu (Birbilis and Buchheit 2005;Cavanaugh, Birbilis et al. 2007), which is evidenced by the reliable corrosion performance of 3xxx commercial Al alloy (Zamin 1981; Seri and Tagashira 1986; Tahani, Chaieb et al. 2003; Liu and Cheng 2011).

Influence of Lithium

The addition of Li in Al alloys is efficient at significantly reducing alloy density whilst increasing strength – making it an obvious contender in the range of transport, namely aero, applications. Al-Li alloys are a rather specialised field that spans the past five decades, with descriptions originally in the 8xxx series compositional space (with principally Li rich compositions). Such so-called 1st generation Al-Li alloys were only used in specialised applications owing to their susceptibility to cracking. The cracking issue was later managed via new alloy compositions and thermomechanical processing (2nd generation Al-Li alloys), however until relatively recently Al-Li alloys were not so popular owing to relatively high corrosion rates and localised forms of corrosion propagation. Most recently, the 3rd generation of Al-Li alloys has gained significant attention and growing usage in commodity aerospace applications. These 3rd generation alloys are actually 2xxx series alloys, with less Li than Cu. These new 2xxx series alloys will be a significant alloy of the future, whilst still further research is required (from a corrosion perspective) to fully understand the performance, particularly as a function of thermomechanical treatment. Some abridged information regarding the role of Li upon corrosion is included here. In Li rich alloys, the formation of strengthening phase, Al$_3$Li which is dispersed homogeneously throughout the matrix, is responsible for the increase in strength (Lavernia and Grant 1987). It is however detrimental to corrosion as Al$_3$Li initially form along the grain boundaries. As Li is an active (i.e. less noble) element, this will localise dissolution to Li rich regions and therefore susceptibility to attack such as intergranular corrosion is high (Martin 1988). When Cu is also added in conjunction with Li (in alloys such as AA2090) the precipitation of phases such as *T1* (Al$_2$CuLi) occurs. There are two modes of attack associated with *T1*, one of which *T1* at the precipitate free zone is dissolved forming small pits, while the other is when *T1* undergo selective dissolution along with dissolution of the adjacent Al matrix leaving larger pits (Buis and Schijve 1992; Buchheit, Wall et al. 1995).

Influence of Activating Elements (I.E. PB, SN)

Lead (Pb) and tin (Sn) are usually present in low levels as trace elements (Gundersen, Aytaç et al. 2004; Premendra, Terryn et al. 2009). When present in trace amounts, their influence is minimal or negligible. When (by say, recycling or contamination) the levels rise above the solubility limits, the presence of Pb leads to segregation that results in Pb-rich film at the metal - oxide interface when the alloy is heat treated at elevated temperature (Sævik, Yu et al. 2005) causing the Al oxide film to destabilise particularly in the presence of chloride. The disruption of Al oxide film leads to an increase in anodic reaction rate which not only increases the pitting susceptibility, but can activate the entire surface. This process is called anodic activation, and has been well documented for a number of years by studies from the group of Nisancioglu (Keuong, Nordlien et al. 2003; Gundersen, Aytaç et al. 2004;Yu, Saevlk et al. 2004; Yu, Sævik et al. 2005; Walmsley, Sævik et al. 2007; Jia, Graver et al. 2008; Graver, Pedersen et al. 2009; Anawati, Graver et al. 2010; Graver, van Helvoort et al. 2010; Anawati, Diplas et al. 2011). There have been some efforts to reduce the activation effect of Pb by addition of more noble alloying elements such as Cu in the hope that the addition of Cu may alter the surface potential - hence reducing the activation (Anawati, Diplas et al. 2011). A similar result was observed for the addition of Mg (Jia, Graver et al. 2008), however, such methods are not viable on the basis that the Pb interfering with the oxide is an effect in addition to any changes in the alloy potential. The presence of Sn along with Pb however reduces the dissolution rate when annealed at the maximum temperature of 600°C for an hour at which Sn is found to dissolve in the aluminium solid solution diluting the Sn concentration on the surface (Graver, Pedersen et al. 2009).

Influence of other Element, Including ZR, CR, SC, TI, W AND SR
These elements are typically added independently in small amounts (i.e. <0.2 wt%) for the purpose of grain refinement, to reduce recrystallisation as well as minimising the effect of intermetallic compounds (Vetrano, Henager Jr et al. 1998). Elements such as Zr and Ti are able to form intermetallics at high temperatures in the Al melt, and persist as finely dispersed particles of Al_3Zr and Al_3Ti within the solidified matrix, which, based on their fine size (i.e. <<1 μm), have a minimal impact on corrosion (Scully, Knight et al. 1993). Similarly with scandium (Sc) additions above the solubility limit the formation of Al_3Sc will occur, contributing to the strength and significantly reducing recrystallisation during thermomechanical processing (Cavanaugh, Birbilis et al. 2007). In general,

and neglecting Al_3Fe, such dispersoids are based on the Al_3X system where X is Zr, Ti, Sc, Er, etc, and taking the form of fine insoluble dispersoids which are functional in grain inoculation and refinement. As such, there are specific alloying additions of Ti and Zr to high strength alloys such as AA7075 (Senkov, Bhat et al. 2005; Zou, Pan et al. 2007; He, Zhang et al. 2010).

Whilst not studied in detail, it has been posited that the ability to suppress recrystallisation leads to lower corrosion rates by avoiding the formation of high angle grain boundaries (Fang, Chen et al. 2009). Furthermore, there are also complex second and third order interactions between sparingly soluble elements that extend beyond the predictions of simple phase diagrams. An example is that the addition of Sr will impact intermetallics such as Al_5FeSi, making them smaller in both their size and volume fraction (Ashtari, Tezuka et al. 2003; Eidhed 2008). Such an effect has a role in corrosion by minimising the number of intermetallic sites. The purpose however of this section, is not to describe the metallurgy, as that has been done classically as far back as Mondolfo (Mondolfo 1971; Mondolfo and Barlock 1975), but to emphasise the microstructures direct impact on corrosion.

In common alloys prepared by conventional casting technologies, transition metals (TMs) such as W, Mo and Cr are not employed owing to their very low solubility limits. However it is important to note that when prepared in sputtered or thin film forms, such Al-TM alloys display the lowest corrosion rates of all the Al-alloys. Shaw successfully produced the alloys by sputter deposition and found that these elements increased the pitting potential and passivity of the alloys as well as inability to form second phase upon heat treatment (Shaw, Fritz et al. 1990; Shaw, Davis et al. 1991). The work of Frankel also showed promising results (Frankel, Russak et al. 1989; Frankel, Davenport et al. 1992; Tailleart, Gauthier et al. 2009).

The Role of Microstructure on Corrosion

In order to understand the corrosion performance of Al alloys, and following on from the previous section, an appreciation of the microstructure is vital. Alloying elements and thermomechanical processing play an important role in dictating the type of microstructure produced. For homogeneous alloys, such as pure Al or 5xxx series alloys, corrosion susceptibility is low due to lack of pre-existing microstructural attack sites. The main concern however is regarding heterogeneous alloys, particularly the higher strength Al alloys such as the 2xxx, 7xxx and heat-treatable 6xxx series, where microstructural heterogeneity is a necessity. The most common features of a microstructure are the intermetallic particles which are classified into precipitates (forming from nucleation and growth, nominally 1nm to 300nm in diameter), constituent particles (from insoluble or impurity elements, unable to redissolve, nominally a few microns, to a few tens of microns, in size) and dispersoids (nominally << 1 micron in size) (Polmear 2006). Each of these features consists of different electrochemical characteristics (including their native electrochemical potential and the currents they can sustain at a given potential characteristic of the alloy which they populate) and act as the sites which dictate the severity of corrosion attack. Work on categorising such intermetallics in relation to corrosion is plentiful and now has been going on for several decades (Mazurkiewicz and Piotrowski 1983; Scully, Knight et al. 1993; Buchheit, Grant et al. 1997; Birbilis and Buchheit 2005; Eckermann, Suter et al. 2008; Goswami, Spanos et al. 2010; Boag, Hughes et al. 2011; Hughes, Boag et al. 2011). We make the distinction in this chapter that whilst such intermetallics are responsible for corrosion initiation steps, including pitting, we will not cover corrosion propagation in detail (i.e. stress corrosion, or intergranular corrosion) since they would require a dedicated chapter.

The knowledge of intermetallic chemistry and the electrochemistry allows a prediction of the mode of corrosion and the propensity of the attack (Cavanaugh, Buchheit et al. 2009). Theoretically, a more active particle (i.e. less noble) will become a local anode and consequently corrode while the more noble particles become cathodes (Szklarska-Smialowska 1999). This is not always a true reflection of kinetics however, as reported by Birbilis where the capability of the element to sustain the cathodic reduction process cannot be the deduced by relative nobility as well (Birbilis and Buchheit 2005). As corrosion occurs upon Al alloys, particularly pitting and early damage accumulation, two types of corrosion mode are identified. In one mode of attack whereis the intermetallic is classified as a cathode, the surrounding matrix tends to corrode leaving a ring shape around the particle or also called trenching.

There is still some uncertainty on whether or not the trench itself is a result of microgalvanic coupling alone, or if the major contributor is local pH elevation, however a good treatise of this topic was given in a multi-part series of papers by the group at Virginia (Ilevbare, Schneider et al. 2004; Schneider, Ilevbare et al. 2004; Schneider, Ilevbare et al. 2007). In some cases damage may propagate to the base of the particle and eventually lead to particle fall out (Buchheit, Grant et al. 1997). The other mode of attack is when the intermetallic acts as anode and matrix as cathode; whereby the intermetallic will corrode leaving a cavity on the surface. Finally, in some microstructurally complex systems with ternary and above alloying additions, another type of attack found to not follow the traditional way of determining anode and cathode is incongruent dissolution commonly found in 2xxx and 7xxx due to the presence of S phase (Al_2CuMg) (Buchheit, Grant et al. 1997; Guillaumin and Mankowski 1998). This intermetallic contains Cu and Mg, whereby (Blanc, Lavelle et al. 1997;Buchheit, Martinez et al. 2000)., S phase experiences selective dissolution of the highly active Mg hence leaving Cu remnants at the bottom of the pit (Buchheit, Grant et al. 1997; Büchler, Watari et al. 2000) or redistrubuted near the particle site. Although the more detailed study of Boag et al. indicated that Al might be preferentially removed in the initial stages of attack on S-phase (first 2.5 minutes) as both Cu and Mg were observed to be present after Al dissolution at 5 minutes. This may have been due to a combination of the types of aluminium hydroxyl-chloro complexes that form and the partial switching of the areas of S-phase to Cu remnants where cathodic reactions lead to the formation of insoluble $Mg(OH)_2$ (Boag, Hughes et al. 2011). After 15 minutes the Mg is removed as well and no chloride was detected on these particles. The attack then continues with the dissolution of Al matrix (Buchheit, Grant et al. 1997; Guillaumin and Mankowski 1998). Localized attack of the intermetallic also influenced by the chloride content and pH of the electrolyte. Higher chloride content is widely reported to have higher pitting occurrence due to passive layer disruption of chloride ions (Seri 1994;Blanc, Lavelle et al. 1997; Ilevbare, Schneider et al. 2004) but this does not necessarily mean the pit will propagate deeper (Cavanaugh, Buchheit et al. 2010). There exists a dedicated monograph on this topic (Muster, Hughes et al. 2009). It has also been noted that the intermetallic Mg_2Si can undergo similar incongruent dissolution, whereby Si enrichment occurs at the expense of dissolving Mg (Birbilis and Buchheit

2005; Jain, 2006; Eckermann, Suter et al. 2008; Gupta, Sukiman et al. 2012).

The revelation of a large number of microstructure vs. corrosion micrographs will not be done herein, however if the readers are interested, a nice expose of such images exists in Cavanaugh (Cavanaugh 2009). Instead, a demonstration is given here. Figure 6 shows the micron scale microstructure for 2024-T3 and 5083-H116 before and after corrosion exposure in 0.1M NaCl for 14 days. These relatively low magnification images do not reveal the precipitate structure in AA2024-T3, instead showing the coarse intermetallics that exist in the alloys. What is observed is that a higher intermetallic density gives more possible sites for localised attack. In the relatively widely studied 2024-T3 (Guillaumin and Mankowski 1998; Schmutz and Frankel 1998; Campestrini, van Westing et al. 2000;Leblanc and Frankel 2002; Boag, Hughes et al. 2009; Hughes, MacRae et al. 2010; Ralston, Birbilis et al. 2010; Boag, Hughes et al. 2011; Hughes, Boag et al. 2011; Zhou, Luo et al. 2012). pitting attack is associated with the Cu containing intermetallic such as *S* phase (coarse and precipitated) and the intermetallic density in AA2024-T3 is high, owing to the very large number of alloying elements which can leave the possibility of forming constituent particles based on Fe, or Si, or Cu, or Mn, etc. In contrast to AA2024, the alloy 5083-H116 with a smaller number of alloying elements, and Cu free, has a lower constituent number density. in addition to the Mg remaining in solid solution and reduces the susceptibility to localised attack (Vetrano, Williford et al. 1997; Aballe, Bethencourt et al. 2001;Yasakau, Zheludkevich et al. 2007).

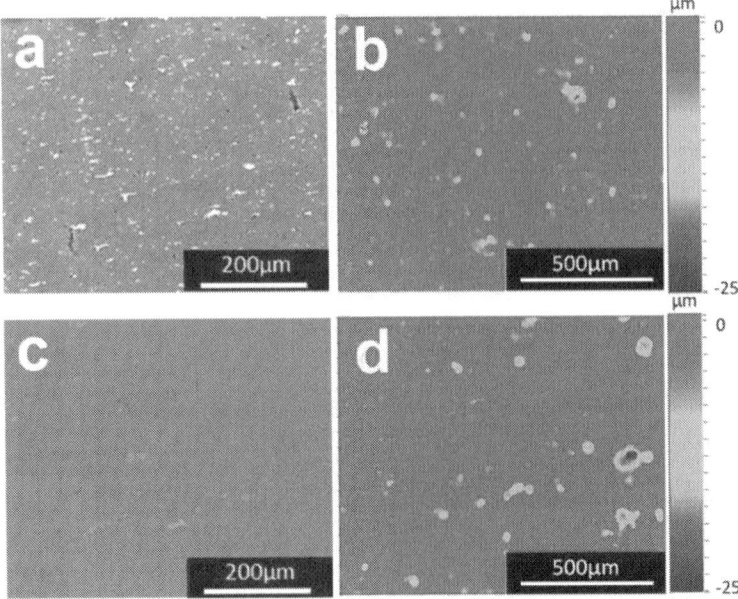

Figure 6: SEM images for AA2024-T3 and AA5083-H116 before exposure in (a) and (c) and after exposure in 0.1M NaCl for 14 days in (b) and (d). These are comparative images to show the extent of damage following immersion.

CORROSION PROTECTION

This section covers general approaches to protection of aluminium alloys in view of recent advances in the understanding of alloy microstructure. It includes an overview of pretreatment processes such as anodising, conversion coating and organic coatings (barrier and inhibitor combinations). It will examine recent advances in inhibitor design such as building in multifunctionality and touch upon self-healing coating systems. Approaches using multifunctionality can target anodic and cathodic reactions more effectively than using individual monofunctional inhibitors.

Standard metal finishing processes, which have been used for many years, are likely to continue to be used into the future unless they contain chemicals that are targeted for replacement such as chromium. The function of these coatings is primarily to provide better adhesion properties for paint coatings and a secondary role is to provide corrosion protection. The general approach for applying these coatings relies on metal finishing treatments (treatment prior to painting involving immersion in acidic and alkaline baths)) with the objective of reducing the heterogeneous nature of the metal surface such as removing the NSDL and second phase particles (Muster 2009). This is achieved in multistep treatment processes for metal protection (Twite and Bierwagen 1998; Buchheit 2003; Muster 2009) as for instance:

- selective deoxidation (IM particle removal and surface etching);
- deposition or growth of a manufactured oxide via electrochemical (anodising) or chemical (conversion coating) means;
- use of an organic coating for specific applications, normally including a primer and a top-coat.

On aluminium, most anodised coating processes produce an outer oxide with a cellular structure on top of a thin barrier layer that provides some protection against corrosion. Inhibitors can be incorporated into the outer porous layer of the anodized layer during formation or as a seal after formation to offer some extra protection upon damage. Chromic acid anodizing is one of a number of processes that are available for electrochemical growth of surface protective oxides. More environmentally friendly alternatives to chromic acid anodizing such as sulfuric, sulfuric-boric, sulphuric-tataric and phosphoric based processes have been available for a long time. There have been a number of recent advances in reducing the energy consumption of anodizing processes as well as improving coating properties. These advances are based on an improved understanding of the alloy microstructure described above and involve selective removal of second phase particles as part of the anodising process.

An alternative approach to anodizing is to precipitate a coating on a surface through chemical means called conversion coatings. For high strength Al-alloys such as 2xxx and 7xxx series chromate conversion coating (CrCC) is still the preferred process. Replacements for the toxic chromate-based conversion coatings include a range of treatments based on self-assembled monolayers, sol-gel chemistries, Ti/Z oxyfluorides, rare earth, cobalt, vanadates, molybdates and permanganate processes (Twite and Bierwagen 1998; Buchheit 2003; Kendig and Buchheit 2003). These processes are widely developed for chemically pretreated surfaces that have nearly all the IM particles removed (by chemical pretreatment) and are not specifically designed to address electrochemical and compositional variations found for a heterogeneous surface such as when the IM phases are present. Work like that in (Buchheit and Birbilis 2010) depicting the reaction rate variation across the surface, however, opens an avenue to start designing inhibitors where the initial reaction rate distribution across a surface can be significantly reduced to limit the overall activity of the surface. In this context reaction of inhibitive phases with manufactured IM compounds as well as IM particles within the alloy have been studied for a number of systems (Juffs, Hughes et al. 2001; Juffs 2002; Juffs, Hughes et al. 2002; Birbilis, Buchheit et al. 2005; Scholes, Hughes et al. 2009).

Once the anodised or conversion coating is applied, the surface is ready to receive the organic coating. There are many different types of organic coatings, however because of the focus on 2xxx and 7xxx alloy used in the aerospace industry this section will only deal with that application area. The organic coating system usually consists of a primer and a topcoat. The primer is the main protective layer including corrosion inhibitors that can be released when corrosive species or water reach the metal. From the perspective of providing protection for the underlying aluminium alloy, the inhibitor needs to be available during a corrosion event at a concentration higher than the minimum concentration at which the inhibitor stops corrosion (critical concentration). While this sounds obvious, the critical inhibitor concentration needs to be maintained over many years for structures such as airframes, where maintenance may not be possible in parts of the aircraft because of poor access. The chromate systems itself provide continuous protection and repair to the surface for as long as the dose of chromate remains above the critical concentration. This mechanism of inhibitor release and metal protection is recognized as a self-healing mechanism, since the release of the active species recovers the protective layer on top of the metal.

The search for green inhibitors as replacements for chromate has been driven by legislative imperatives for a number of years. Needless to say, replacement inhibitors do not have the same intrinsic inhibitive power at low solubility as chromate. Thus solubility, inhibitive power and transport within the primer system (which consists of a number of inorganic phases as well as the epoxy) ultimately mean that finding a replacement for chromate is difficult. This means that alternatives must be present at higher concentration leading to the use of more soluble compounds and consequently encapsulation as a method of regulating the response to external or internal triggers emerges as a prospective way to achieve this objective.

Many current inhibitors are water soluble salts and thus ionic. Consequently, they exist as either anions or cation in solution and perform the single function of anodic or cathodic inhibition. So the simplest improvement to inhibitor design is to increase the functionality by finding compounds which play both a cationic and anionic inhibitive role. A large range of cations including Zn, Ca, and rare earths (Bohm, McMurray et al. 2001; Du, Damron et al. 2001; Kendig and Buchheit 2003; Taylor and Chambers 2008; Muster, Hughes et al. 2009) have been combined with either organic (Osborne, Blohowiak et al. 2001; Sinko 2001; Voevodin, Balbyshev et al. 2003; Khramov, Voevodin et al. 2004; Blin, Koutsoukos et al. 2007; Taylor and Chambers 2008; Muster, Hughes et al. 2009) or inorganic (oxyanions, carbonates, phosphates, phosphites, nitrates, nitrites, silicate (Bohm, McMurray et al. 2001; Sinko 2001; Blin, Koutsoukos et al. 2007; Taylor and Chambers 2008) compounds.

Anions with dual functionality, such as some of the transition metal oxyanions which are both oxidants and anions, have been investigated extensively. The oxidizing oyxanions or some organophosphates have some degree of bio-inhibition required for some applications. Substitution of different organophosphates into rare earth-based inhibitors provide versatility in designing inhibitors for specific applications (Birbilis, Buchheit et al. 2005; Hinton, Dubrule et al. 2006; Ho, Brack et al. 2006; Blin, Koutsoukos et al. 2007; Markley, Forsyth et al. 2007; Markley, Hughes et al. 2007; Forsyth, Markley et al. 2008; Deacon, Forsyth et al. 2009; Scholes, Hughes et al. 2009). Thus Ce(di-butyl phosphate)$_3$ is a good inhibitor and relatively "green" whereas Ce(di-phenyl phosphate)$_3$ is also a good inhibitor, but the diphenyl phosphate also has strong bio-inhibition characteristics (García 2011). However, good bio-inhibition usually means that there are increased environmental and health risks. Obviously the number of cathodic and anodic inhibitors means that there are an enormous number of possible combinations, particularly if ternary and quaternary combinations are considered. Hence high-throughput techniques are being used to assess new inhibitor.

As pointed out above, the kinetics of inhibitor release are of the utmost importance since the inhibitor should be available at levels above the critical inhibitor concentration. Optimization of the release kinetics by novel delivery systems becomes integral to incorporation of new inhibitors.

There are a number of different mechanisms investigated for release of healing agents or corrosion inhibitors which can be incorporated into organic coatings. Both mechanical damage and water are triggers for inhibitor release. In the former case mechanical damage breaks capsules containing water soluble inhibitors. In the latter case water dissolves inhibitor directly incorporated in the primer. Droplet formation within defects such as scratches means that the inhibitor is only released when required i.e., when the defect is moist (Furman, Scholes et al. 2006). There is some evidence to suggest that initial high release of inhibitors may be facilitated through atmospheric exposure of the intact paint where penetration of water into the film "prepares" the inhibitor, probably via surface hydrolysis reactions, within the paint, for diffusion and release into the defect (Joshua Du, Damron et al. 2001; Furman, Scholes et al. 2006; Scholes, Furman et al. 2006; Souto, González-García et al. 2010). The presence of water in the film allows soluble inhibitor species to be released into the paint system and diffuse to the metal/coating interface to provide in-situ corrosion prevention or repair called pre-emptive healing (Zin, Howard et al. 1998; Osborne, Blohowiak et al. 2001; Mardel, Garcia et al. 2011). Thus it has been demonstrated that water can trigger cerium dibuthylphosphate (Ce(dbp)$_3$) release into an epoxy matrix resulting in improved adhesion and resistance to filiform corrosion attack through interfacial modification (Mardel, Garcia et al. 2011)

In terms of delivery systems, hard capsules, which have been used in polymer healing (Dry 1996;White, Sottos et al. 2001; Mookhoek, Mayo et al. 2010) need to be smaller for paint systems particularly in the aerospace industry where coatings are typically 20 μm or less (Yin, Rong et al. 2007;Fischer 2010; Hughes, Coles et al. 2010; Mookhoek, Mayo et al. 2010). In polymer applications, capsules up to a few hundred microns can be accommodated (Yin, Rong et al. 2007; Wu, Meure et al. 2008; Tedim, Poznyak et al. 2010). The concept of encapsulation has already been successfully applied to protective organic coatings under different concepts: i) liquids filling completely the void created by the damage by adopting a bi-component systems where one component is encapsulated and the other distributed in the matrix (Cho, White et al. 2009), or single based components with water reactive oils like linseed and tung oils (Suryanarayana, Rao et al. 2008; Samadzadeh, Boura et al. 2010)and ii) liquids (i.e. silyl esters) forming a hydrophobic and highly adhesive layer covering the metallic surface by reaction with the underlying metal and the humidity in air (García, Fischer et al. 2011). One adaption for capsules is to increase the volume of self-healing material by manufacturing rods instead of spheres. Rods with the same cross-sections as spheres can deliver larger volumes of material (Bon, Mookhoek et al. 2007; Mookhoek, Fischer et al. 2009). For inhibitors, their role is to prevent a surface reaction (corrosion) and therefore, the volume of material required is much smaller than that required to actually fill the defect. Consequently, there has been considerable effort looking at "nano-containers" (Voevodin, Balbyshev et al. 2003; Raps, Hack et al. 2009; Tedim, Poznyak et al. 2010).

Water is the most obvious trigger since it can permeate most polymers. pH variations are more specific and respond to the pH excursions that occur in corrosion reactions and by an understanding reactions that occur at different sites in the alloy microstructure. The presence of chloride ions (and other anions) within the coating can be used as specific triggers for the release of corrosion inhibitors and uptake of corrodents using anion exchange materials, such as layered double hydroxides (e.g. hydrotalcites) (Tedim, Poznyak et al. 2010) (Bohm, McMurray et al. 2001; Buchheit, Guan et al. 2003;Williams and McMurray 2003; Zheludkevich, Salvado et al. 2005; Mahajanarn and Buchheit 2008). In this context hydrotalcites have been loaded with vanadate, chromate, nitrate and carbonate which exchange for chloride ions and prevent interfacial damage (Bohm, McMurray et al. 2001; Williams and McMurray 2003; Mahajanarn and Buchheit 2008). The incorporation of Mg particles into paint act as sacrificial anodes to protect Al alloys and steels (Battocchi, Simoes et al. 2006).

RECENT ADVANCES IN ASPECTS RELATED TO CORROSION OF ALUMINIUM ALLOYS

The search for new multifunctional inhibitors has led to the development of high throughput and combinatorial assessment of new combination of inhibitors. These include multielectrode techniques, and high throughput versions of standard corrosion tests. A range of new electrochemical techniques including AC/DC/AC, SVET, LEIS, SECM and SIET will also be described.

Since corrosion of aluminium alloys tends to be dominated by electrochemical processes, most of the techniques employed for the evaluation of corrosion and protection are based on electrochemical approaches. Furthermore, combining electrochemical techniques with other microscopic, analytical and spectroscopic techniques enables the identification of corrosion products in solution (such as inductive coupled plasma (ICP) and UV-Vis). This combination provides an even broader mechanistic understanding of the level of corrosion and/or corrosion protection.

The increasing number of corrosion inhibitor alternatives to chromates has boosted interest in developing high-throughput techniques and combinatorial assessment of new corrosion inhibitors in aqueous solution. At the same time, the traditional techniques (accelerated or not) employed in the evaluation of the performance of organic coatings require long evaluation periods and are relatively expensive to run, and only offer qualitative or semi-quantitative information at best (e.g. salt fog spray tests). For these reasons, new accelerated techniques for the evaluation of coating performance that offer quantitative results are needed.

Figure 7 shows a simplified flowchart for the formulation of anticorrosive (organic) coatings. The chart includes some of the most common techniques employed in corrosion inhibitor and coating performance evaluation. In the figure, the parallel and complex line of the development of the polymeric matrix (i.e. organic coating) is not included, but awareness of its existence is important, since factors such as the corrosion inhibitor/coating matrix compatibility should be taken into account. For the development of anticorrosive organic coatings, several steps are proposed:

Formulation of Inhibitors:

The number of corrosion inhibitor candidates is virtually unlimited, and is motivated by the urgent need to replace chromate based inhibitors by environmentally friendly and non-toxic ones, as well as the development of new concepts such as self-healing and synergies between anodic and cathodic inhibitors which open up the broad range of possibilities of organic chemistry. One example of the complexity of the introduction of organic compounds as corrosion inhibitors is the effect that the position of certain groups in a cyclic organic compound can have in the corrosion protection efficiency (Harvey, Hardin et al. 2011).

Evaluation in Aqueous Solution:

Once the inhibitors have been formulated, they can be tested by means of traditional aqueous solution tests such as electrochemical impedance spectroscopy (EIS), potentiodynamic polarisation (PP), immersion tests and weight loss/gain. Also local electrochemical techniques (see point 4- evalaution of organic coatings) give very important information of the mechanisms of corrosion protection offered by the different species in solution. Since traditional techniques require long periods of time and a large number of samples, the introduction of high- throughput techniques as a preceding step is important in order to reduce the number of inhibitors that enter further evaluation processes using traditional aqueous solution tests. It is necessary to highlight that high-throughput techniques are not aimed at replacing traditional tests but at complementing them in order to reduce cost and time in the corrosion inhibitor selection process. Some examples are: single metal, multielectrode array (Chambers, Taylor et al. 2005; Chambers and Taylor 2007), microchannels (White, Hughes et al. 2009) and multi metal multielectrode (Muster, Hughes et al. 2009; García, Muster et al. 2010; Kallip, Bastos et al. 2010).

Introduction into an Organic Matrix:

The introduction of pigments into organic coatings adds some extra difficulties to the whole process, leading to a lot of extra research to avoid undesirable reactions between the polymer matrix and the inhibitors. Some of the parameters to take into account are the ratio between the pigment volume concentration (PVC) and the critical pigment volume concentration (CPVC), the possible side reactions between the polymer and the pigment, with consequences in parameters like the barrier properties, gloss, active corrosion protection, and adhesion amongst others. At the same time, parameters such as contaminant reduction and parameters related to the polymeric matrix itself (such as adhesion and the glass transition temperature (Tg)) should be considered. In any case, once the pigments have been introduced into the organic coating, the coating's performance has to be tested for protection efficiency and if results are promising, then start the optimization process.

Evaluation of the Performance of Organic Coatings:

As in the case of aqueous solution tests for evaluation of corrosion inhibitors, the introduction of accelerated tests to evaluate coatings performance is necessary to reduce the amount of time and number of samples that move into traditional assessment. Several techniques have been proposed in this direction, such as the technique AC/DC/AC (Hollaender 1997; Bethencourt, Botana et al. 2004;Rodriguez, Gracenea et al. 2004; Garcia and Suay 2006; Garcia and Suay 2006) (García, Rodríguez et al. 2007; Garcia and Suay 2007) (Garcia and Suay 2007; García and Suay 2007; García and Suay 2009), (Poelman, Olivier et al. 2005; Allahar, Bierwagen et al. 2010; Allahar, Wang et al. 2010), and thermal cycling (Bierwagen, He et al. 2000). Also new concepts like the use of flow induced degradation (Wang 2009) are interesting for developing accelerated testing techniques.

Optimisation:

Before entering the pre-commercialization phase, the final step of anticorrosive coatings formulation is the optimization, which is the improvement of the system by modifying pigments concentration, type, and delivery systems to improve and extend the service lifetime protection and compatibility with the matrix. This step is iterative as shown in Figure 7. The whole process from conception to commercialization of the system can take several years.

Due to the impossibility to cover the broad amount of existing (new) high-throughput techniques for selection of corrosion inhibitors for aluminium alloys and accelerated tests for evaluation of protective organic coatings on aluminium alloys, we focus on two electrochemical techniques that have attracted a broad interest most recently due to their high potential and relatively well understood evaluation procedure.

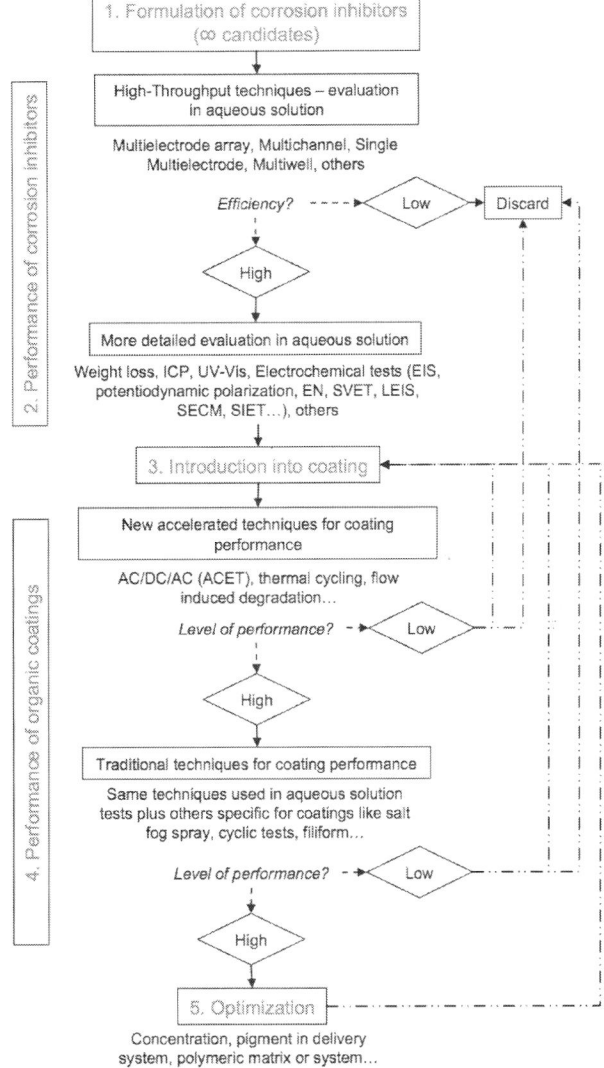

Figure 7: Simplified flowchart for anticorrosive coatings development departing from the inhibitor design or formulation.

HIGH THROUGHPUT ASSESSMENT

In terms of high-throughput techniques, those based on electrochemical approaches are of most interest, since their measurement principles are closely related to familiar research techniques such as Electrochemical Impedance Spectroscopy (EIS) and potentiodynamic polarisation (PP), while at the same time offer quantitative information about the corrosion and corrosion protection mechanisms.

Chambers and Taylor (Chambers, Taylor et al. 2005; Chambers and Taylor 2007) first presented the use of multi-electrode arrays for rapid screening of corrosion inhibitors in different concentrations and pHs, putting identical pairs of AA2024-T3 wires into a large number of separate reaction cells containing different inhibitive solutions and applying a 100mV potential between the two wires in each cell (Figure 8(a)). The current between both electrodes was used to determine the polarisation resistance. With this set-up they were able to evaluate for one metal many different combinations of inhibitors and their synergistic behavior, while at the same time showing its possible application for evaluating the influence of the pH on inhibitors' efficiency.

Based on the work of Taylor and Chambers, Muster et al. (Muster, Hughes et al. 2009) proposed a variation of the method using a combination of different pairs of metals assembled together (Figure 8(b)) to form what was presented as a multielectrode (ME). The basic idea of this ME was to test a combination of nine pairs (or as many as are interesting) of different metals in the same electrode configuration connected by means of a multiplexer to a potentiostat/galvanostat. Measurements consist of applying 100mV between a selected pair of the same metals within the ME and measure the current flow between them, repeating afterwards the same procedure for the other metal pairs. This set up was employed to rapidly evaluate, without removing the ME from the solution, the concentration range at which a particular inhibitor or combination of inhibitors were offering corrosion protection. Hence, the setup allowed the determination of optimal metal-inhibitor combinations, while significantly reducing the evaluation time with respect to conventional PP tests, without the need of a reference electrode, which simplifies the experimental setup.

In a second paper (Garcia, Muster et al. 2010), the authors studied the effect of the pH and inhibitor type on the correlation between the ME and PP for AA2024-T3. The findings were promising due to the high level of correlation between the ME and traditional techniques, although some discrepancies were found for corrosion inhibitors that can speciate or precipitate at certain pHs. Nevertheless, the non- correlation was assumed to be dependent on the type of corrosion inhibitors and not due to conceptual or experimental mismatching between techniques.

A second concern with the ME was the possibility of cross-contamination due to the presence of several metals in one solution. Garcia et al. (Garcia, Muster et al. 2010) also addressed this problem studying the effect of cross-contamination for the AA2024-T3 wires within the ME. This study showed that if there was cross-contamination then it was not significant enough to influence the results. Despite these results, some more studies should be performed with the ME to check cross-contaminations for other metals such as AA7075-T6 which could be more susceptible to copper plating coming from other metals such as AA2024-T3.

The results obtained so far with multielectrodes (multielectrode array and ME) are very promising and relatively extended information can be found in literature. Nevertheless, more studies and data treatment simplification need to be performed to completely validate these techniques and lead them to an industrial application level.

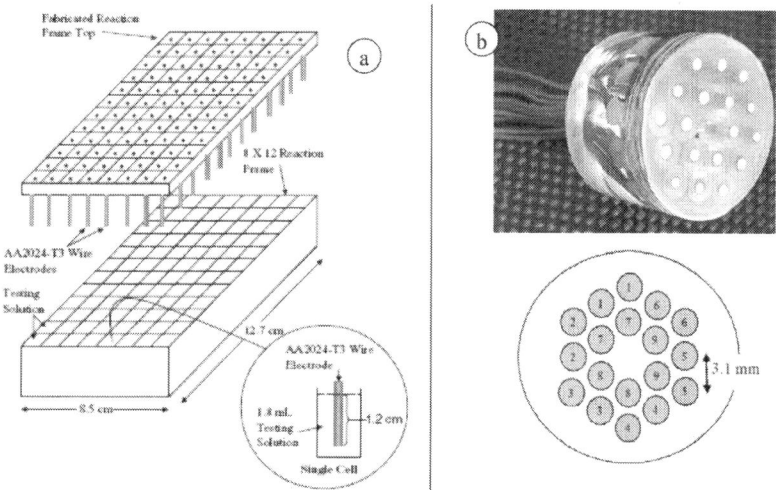

Figure 8: a) single metal multiarray (Chambers and Taylor 2007); b) multimetal multielectrode (Muster, Hughes et al. 2009)

AC/DC/AC ACCELERATED TECHNIQUE FOR COATING EVALUATION

This technique has already reached its maturity and is actually employed at an industrial level under the name of accelerated electrochemical technique (ACET) (Medco). Although the industrial application differs from the research technique (AC/DC/AC), the evaluation and testing are conceptually similar. The early version of the AC/DC/AC technique was performed for the packaging industry (Hollaender 1997). The accelerated technique developed by Hollaender (named AC/DC/AC) was based on the use of temporary stresses to accelerate degradation, and consisted of a first EIS test (AC) to evaluate the initial state of the coating, followed by a cathodic polarisation (DC) and a new EIS (AC) to detect the degradation of the lacquer.

The work initiated by Hollaender was further developed by Suay, Garcia and Rodriguez who successfully applied a modified version of the AC/DC/AC technique to evaluate the performance of organic coatings for carbon steel protection and compared the obtained results with EIS, salt-fog spray, and cyclic tests. The technique was then tested for liquid paints (Bethencourt, Botana et al. 2004; Rodríguez, Gracenea et al. 2004), powder coatings (García and Suay 2006; García and Suay 2006; García and Suay 2007; García and Suay 2007) and cataphoretic paints to optimize parameters such as cataphoretic potential and curing time (Poelman, Olivier et al. 2005; García, Rodríguez et al. 2007; García and Suay 2007; García and Suay 2009). The new version of the technique included a crucial step: the relaxation of potentials (open circuit potential relaxation) after the application of each cathodic polarisation. Furthermore, the potentials applied during the cathodic polarisation depended on the type of coating that was studied, although -4V for 20 minutes was preferred. Figure 9(a) shows a schematic of the AC/DC/AC technique procedure, including the relaxation step proposed by Garcia and Suay. The AC/DC/AC cycle is repeated 6 times (6 cycles) leading to a testing time per sample of around 24 hours, which is a significant improvement when compared to traditional EIS and salt-fog tests which require weeks or months. Nevertheless, depending on the quality of the coatings the number of cycles could be increased and the relaxation time reduced or extended.

The cathodic polarisation aims to degrade the coating and coating-metal interface (e.g. pore formation and delamination) due to hydrogen and OH− production. If the coating is good then it has a higher number of cycles to degrade and the effects in the impedance and potentials relaxation are less pronounced, while a lower quality coating will display a faster degradation. An example of these effects compared to traditional EIS and salt-fog spray is presented by Garcia et al. (García and Suay 2009)

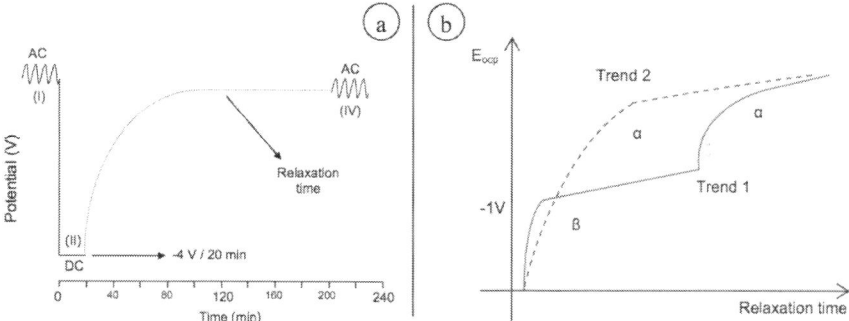

Figure 9: (a) AC/DC/AC testing scheme (García and Suay 2009); (b) Trends in relaxation of potentials.

Apart from the EIS spectral evolution with cycles, for which a higher drop of impedance is related to higher degradation, the evolution of the open circuit potential after polarisation (relaxation of potentials with time) provides extra crucial information about the degradation of the systems under study (García and Suay 2006). When the cathodic polarisation finishes, the potential of the system relaxes leading to two types of trends depending on the quality of the film (Figure 9(b)):

Strongly degraded systems (Trend 1, Figure 9(b)), show two time relaxations of the potential, namely α and β. The first relaxation in time (β) is related to the end of the cathodic reactions that took place at the metallic surface. This relaxation is observed as a quick relaxation around −1V (with small variations depending on the system). The second relaxation (α), which occurs later in time, corresponds to ions and electrolyte leaving the coating. The relaxation (β) could not be detected in all cases due to extremely long relaxation times, although if enough time was allowed this relaxation was then detected, supporting the idea of the two relaxation processes (García and Suay 2009)

Less degraded systems (Trend 2, Figure 9(b)), only show relaxation α which is detected at relatively short times of relaxation. This relaxation will take place at longer times as ions and electrolyte penetrate deeper into the film. More recently, Allahar et al. (Allahar, Wang et al. 2009; Allahar, Upadhyay et al. 2010; Wang, Battocchi et al. 2010) have performed an extensive study on the understanding of the AC/DC/AC technique and its principles, leading to a broader understanding of the relaxation of potentials and supporting previous theories, while at the same time further validating the technique.

In this section we have highlighted the complexity of the selection of corrosion inhibitors and anticorrosive coatings formulation leading to the design and validation of high-throughput and accelerated techniques, which at a certain stage can become commercial techniques. The interest in developing new techniques and understanding those already existing is indeed growing in recent years due to the need of developing environmentally friendly and non-toxic systems.

STAIRCASE IMPEDANCE

Staircase potentio-electrochemical impedance spectroscopy (SPEIS) is based on EIS, nowadays common for investigating electrochemical and corrosion system. The basis of EIS is by measuring impedance over a range of frequency and the data obtained is expressed graphically in Bode or Nyquist plot format. As for SPEIS, it is designed for impedance measurements over a range of frequency as well as over a range of voltages (similar to the signal sequence used for Mott-Schottky analysis (Barsoukov and Macdonald 2005; Orazem and Tribollet 2008). The potential range is divided into several potential steps and each step contains a DC relaxation period for a given rest time (T_R) to allow the current to stabilize; subsequently followed by an EIS test (Zhou, Birbilis et al. 2010). During the DC potential signal, the current at each potential is also recorded. Figure 10 demonstrates this process.

This method is used to assess the kinetic stability of electrochemical reactions over a range of potentials and pH, which gives a rather detailed insight into the corrosion behavior. In SPEIS however, by setting a range of voltage allows more details observation of how the system response at a particular voltage or 'step', even though it is not sufficient to explain the kinetic processes occurring. As reported by Zhou, SPEIS is able to illustrate the effect of pH and potential on the corrosion kinetics with clarity and able to obtain more information beyond the Pourbaix diagram (Zhou, Birbilis et al. 2010), including structural effects such as alloying, etc.

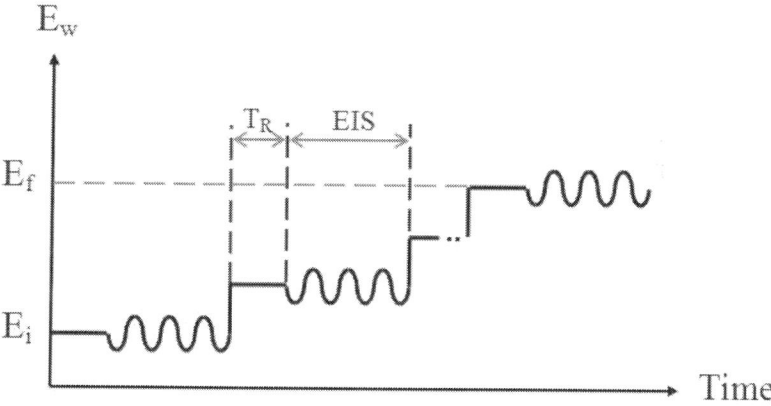

Figure 10: Illustration of polarizing signal during SPEIS

POTENTIOSTATIC TRANSIENTS FOR DETERMINATION OF METASTABLE AND STABLE PITTING

Since pitting is the most common type of corrosion in Al alloys, researchers commonly default to the investigation of pitting potential, E_{pit} as a means for evaluating the corrosion performance. A more noble E_{pit} is often accepted as better resistance to pitting. However, there are some significant limitations in such simple assessments (Gupta, Sukiman et al. 2012). For example, the pitting potential yields no information regarding how many pits form or how large pits may be. Additionally, the environment plays a key role in the severity of pitting damage (i.e. pit depth, pit size) (Cavanaugh 2009).

Pits that form at pitting potential are known as stable pits. However, at potentials slightly below pits stabilization (at E_{pit}), pitting like events occur that are known as metastable pit (small surges of current that repassivate). Wu suggested that individual metastable pitting events influence subsequent events (Wu, Scully et al. 1997), and such events are readily observed in Al-alloys. Figure 11 indicates the metastable pitting region in a typical potentiodynamic polarisation curve, where E_{pit} is also depicted.

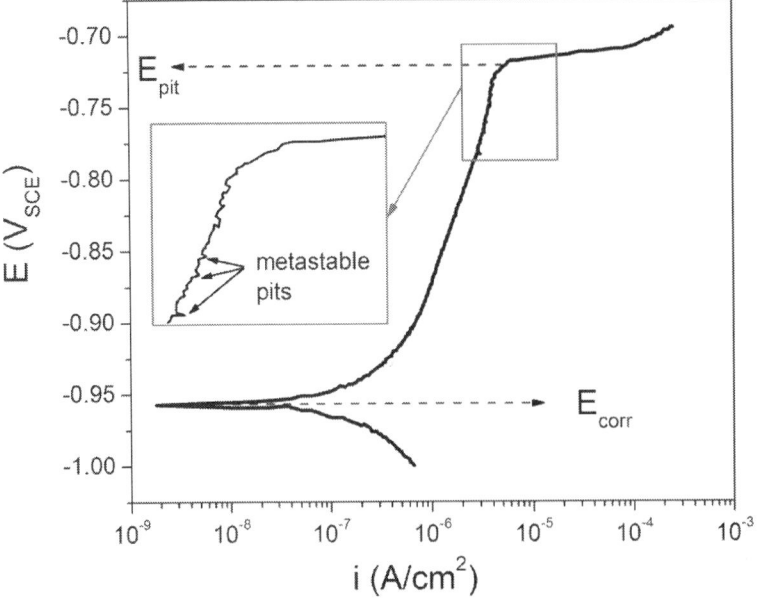

Figure 11: Polarisation curve indicating the corrosion potential E_{corr} and E_{pit} with metastable pits region prior to E_{pit} highlighted in the inset (adapted from (Cavanaugh 2009))

In order to measure metastable pitting, a potentiostatic value is selected at a potential just below E_{pit} where the metastable pitting events are the most frequent (while still in the passive region) (Cavanaugh 2009). Metastable pitting events are measured by counting the current fluctuations when an alloy is held potentiostatically for a period of time.

The transient currents as shown in Figure 12 depend on the nucleation, growth and repassivation of the metastable pits.

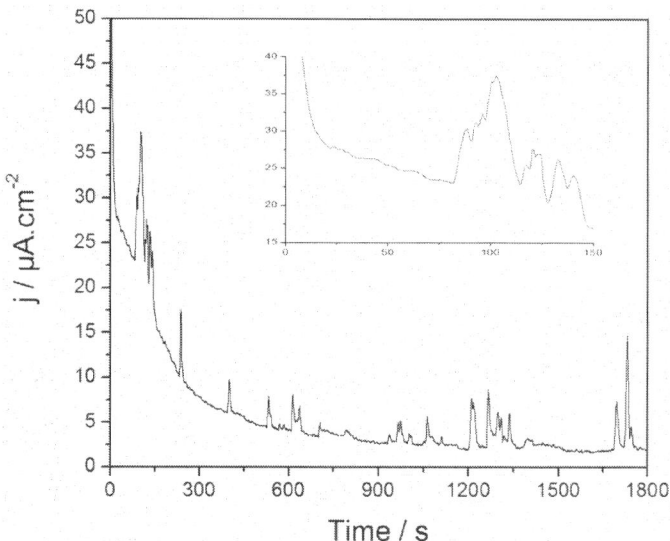

Figure 12: Current transient for AA5083-H116 obtained at 25mV below E_{pit}. Inset shows a zoom of a region of interest that represents the typical transient features (adapted from(Gupta, Sukiman et al. 2012))

The number of metastable pitting is expected to be proportional to the amount of pits formed (Williams, Stewart et al. 1994; Cavanaugh, Birbilis et al. 2012). This is confirmed by (Gupta, Sukiman et al. 2012), where the numbers of metastable pitting from potentiostatic test for various commercial Al alloys correlate with stable pitting tested in a long term immersion test. Ilevbare and Burstein however, stated that multiple metastable events may correspond to only one pit site (Burstein, Liu et al. 2004; Ilevbare, Schneider et al. 2004; Sasaki and Isaacs 2004; Trueman 2005; Speckert and Burstein 2011), however most of such analyses were not performed on Al-alloys. Since there are few studies on metastable pitting of Al alloy, the associated theory and principles will evolve in the coming decade. It is however obvious that metastable pitting analysis can be used to compare the pitting susceptibility between different environments [Cavanaugh 2009] and alloy systems [Gupta, Sukiman et al. 2012] more effectively than an examination of (E_{pit}) alone.

Aside from the methods described above, the use of modern tests does not exclude the use of other traditional tests that require longer testing times and samples such as salt fog spray tests, weight determination, outdoor exposure, or cyclic salt fog-climatic chamber tests, but they aim at providing unique insights that are relevant to specific problems at hand such as pitting germination and inhibitor selection.

It is also worth mentioning here, the great interest that local electrochemical techniques have recently attracted to evaluate the corrosion protection offered when the coatings are damaged (i.e. self-healing). These techniques have been successfully applied to several conceptually different self-healing systems like scanning vibrating electrode technique (SVET) for encapsulated agents (Hughes, Cole et al. 2010;García, Fischer et al. 2011), scanning electrochemical microscope (SECM) for shape memory polymers, selective ion electrode technique (SIET) to evaluate activity of metals (Lamaka, Karavai et al. 2008), and local electrochemical impedance spectroscopy (LEIS) for evaluation of corrosion protection by inhibitor release from coatings (Jorcin, Aragon et al. 2006)

SUMMARY AND CHALLENGES

This chapter has given an abridged and focused treatment of the corrosion behavior of Al alloys on the basis of chemistry, microstructure and environment. Understanding these parameters is crucial in deployment of Al-alloys, but also in the development of more durable Al-alloys. It is obvious that alloying (i.e. chemistry) and microstructure dictates not only mechanical strength – but also corrosion performance. In this chapter we stopped short of discussing corrosion propagation such as intergranular corrosion, exfoliation corrosion and stress corrosion cracking – given space constraints, however such topics (along with hydrogen embrittlement) remain critical for structural alloy deployment.

In regards to corrosion protection, we have attempted to cover some modern developments and present some new techniques to assess the corrosion behavior designed to meet the more complicated and challenging requirements of inhibitor selection / chromate replacement. The information gathered from these techniques is beneficial for future protection and alloy development.

Looking to the future, the current strategy to increase strength by precipitation hardening of a crystalline matrix may no longer be feasible if an alloy with strength higher than 1000MPa is required. Methods to meet this threshold include (the un-upscalable) severe plastic deformation method such as High Pressure Torsion (Liddicoat, Liao et al. 2010) – however the thermal stability of these structures is low. Alternatively the strength of Al based metals could be significantly enhanced (up to 1500MPa) when rapidly quenched to form amorphous alloys (Masumoto 1994; Inoue and Takeuchi 2004; Li, Li et al. 2009; Yang, Yao et al. 2009). Such alloys typically use transition metals such as Zr, Ti, Nb, La, etc (Inoue, Gook et al. 1995; Zawrah and Shaw 2003; Rizzi and Battezzati 2004; Samanta, Manna et al. 2007; Huang, Li et al. 2008; Li, Li et al. 2009), however, again – the stability of such structures is unknown, and the ability form large components is also a challenge. From a corrosion point of view, amorphous Al-alloys appear promising (Manna, Chattopadhyay et al. 2004; Lucente and Scully 2007; Lucente and Scully 2008; Tailleart, Huang et al. 2012) with the corrosion behavior of such alloys is not widely explored therefore leaves a lot of opportunities for new discoveries.

REFERENCES

1. Aballe, A., M. Bethencourt, et al. (2001). "Localized alkaline corrosion of alloy AA5083 in neutral 3.5% NaCl solution." Corrosion Science 43(9): 1657-1674.
2. Allahar, K. N., V. Upadhyay, et al. (2010). "Characterizing the relaxation of the open-circuit potential during an AC-DC-AC accelerated test." Corrosion 66(9): 0950011-09500111.
3. Allahar, K. N., D. Wang, et al. (2009). Real time monitoring of an air force topcoat/mg-rich primer system in b117 exposure by -embedded electrodes.
4. Ambat, R., A. J. Davenport, et al. (2006). "Effect of iron-containing intermetallic particles on the corrosion behaviour of aluminium." Corrosion Science 48(11): 3455-3471.
5. Ambat, R. and E. S. Dwarakadasa (1992). "The Influence of PH on the Corrosion of Medium Strength Aerospace Alloy-8090, Alloy-2091 And Alloy-2014." Corrosion Science 33(5): 681-690.

6. Anawati, S. Diplas, et al. (2011). "Effect of copper on anodic activity of aluminum-lead model alloy in chloride solution." Journal of The Electrochemical Society 158(5): C158-C163.
7. Anawati, S. Diplas, et al. (2011). "Surface Characterization of Heat Treated AlPbCu Model Alloys." Journal of The Electrochemical Society 158(6): C178-C184.
8. Anawati, B. Graver, et al. (2010). "Multilayer corrosion of aluminum activated by lead." Journal of The Electrochemical Society 157(10): C313-C320.
9. Andreatta, F., H. Terryn, et al. (2004). "Corrosion behaviour of different tempers of AA7075 aluminium alloy." Electrochimica Acta 49(17–18): 2851-2862.
10. Ashtari, P., H. Tezuka, et al. (2003). "Influence of Sr and Mn additions on intermetallic compound morphologies in Al-Si-Cu-Fe cast alloys." Materials Transactions 44(12): 2611-2616.
11. Baer, D. R., C. F. Windisch, et al. (2000). "Influence of Mg on the corrosion of Al." Journal of Vacuum Science & Technology a-Vacuum Surfaces and Films 18(1): 131-136.
12. Barsoukov, E. and J. R. Macdonald (2005). Solid State Devices, in Impedance spectroscopy: theory, experiment, and applications, Wiley-Interscience, Inc.
13. Battocchi, D., A. M. Simoes, et al. (2006). "Comparison of testing solutions on the protection of Al-alloys using a Mg-rich primer." Corrosion Science 48(8): 2226-2240.
14. Bethencourt, M., F. J. Botana, et al. (2004). "Lifetime prediction of waterborne acrylic paints with the AC-DC-AC method." Progress in Organic Coatings 49(3): 275-281.
15. Birbilis, N. and R. G. Buchheit (2005). "Electrochemical Characteristics of Intermetallic Phases in Aluminum Alloys." Journal of The Electrochemical Society 152(4): B140.
16. Birbilis, N., R. G. Buchheit, et al. (2005). "Inhibition of AA2024-T3 on a phase-by-phase basis using an environmentally benign inhibitor, cerium dibutyl phosphate." Electrochemical and Solid State Letters 8(11): C180-C183.
17. Birbilis, N., M. K. Cavanaugh, et al. (2006). "Electrochemical behavior and localized corrosion associated with Al7Cu2Fe particles in aluminum alloy 7075-T651." Corrosion Science 48(12): 4202-4215.
18. Birbilis, N., T. Muster, et al. (2011). Corrosion of Aluminum Alloys. Corrosion Mechanisms in Theory and Practice, Third Edition, CRC Press: 705-736.
19. Blanc, C., B. Lavelle, et al. (1997). "The role of precipitates enriched with copper on the susceptibility to pitting corrosion of the 2024 aluminium alloy." Corrosion Science 39(3): 495-510.
20. Blin, F., P. Koutsoukos, et al. (2007). "The corrosion inhibition mechanism of new rare earth cinnamate compounds - Electrochemical studies." Electrochimica Acta 52(21): 6212-6220.

21. Boag, A., A. E. Hughes, et al. (2011). "Corrosion of AA2024-T3 Part I: Localised corrosion of isolated IM particles." Corrosion Science 53(1): 17-26.
22. Boag, A., A. E. Hughes, et al. (2009). "How complex is the microstructure of AA2024-T3?" Corrosion Science 51(8): 1565-1568.
23. Bohm, S., H. N. McMurray, et al. (2001). "Novel environment friendly corrosion inhibitor pigments based on naturally occurring clay minerals." Materials and Corrosion-Werkstoffe Und Korrosion 52(12): 896-903.
24. Bon, S. A. F., S. D. Mookhoek, et al. (2007). "Route to stable non-spherical emulsion droplets." European Polymer Journal 43(11): 4839-4842.
25. Brunner, J. G., J. May, et al. (2010). "Localized corrosion of ultrafine-grained Al–Mg model alloys." Electrochimica Acta 55(6): 1966-1970.
26. Buchheit, R. G. (1995). "A Compilation of Corrosion Potentials Reported for Intermetallic Phases in Aluminum-Alloys." Journal of The Electrochemical Society 142(11): 3994-3996.
27. Buchheit, R. G. (2000). "Electrochemistry of θ (Al2Cu), S (Al2CuMg) and T1 (Al2CuLi) and localized corrosion and environment assisted cracking in high strength Al alloys." Materials Science Forum 331: II/.
28. Buchheit, R. G. and N. Birbilis (2010). "Electrochemical microscopy: An approach for understanding localized corrosion in microstructurally complex metallic alloys." Electrochimica Acta 55(27): 7853-7859.
29. Buchheit, R. G., R. P. Grant, et al. (1997). "Local Dissolution Phenomena Associated with S Phase (Al2CuMg) Particles in Aluminum Alloy 2024-T3." Journal of The Electrochemical Society 144(8): 2621-2628.
30. Buchheit, R. G., H. Guan, et al. (2003). "Active corrosion protection and corrosion sensing in chromate-free organic coatings." Progress in Organic Coatings 47(3-4): 174-182.
31. Buchheit, R. G., Hughes, A.E. (2003). Chromate and Chromate-Free Coatings. Corrosion: Fundamentals, Testing and Protection. C. Moosbrugger. Mterials Park, Oh, USA, ASM International. 13A: 720 -735.
32. Buchheit, R. G., M. A. Martinez, et al. (2000). "Evidence for Cu Ion Formation by Dissolution and Dealloying the Al2CuMg Intermetallic Compound in Rotating Ring-Disk Collection Experiments." Journal of The Electrochemical Society 147(1): 119-124.
33. Buchheit, R. G., L. P. Montes, et al. (1999). "The Electrochemical Characteristics of Bulk-Synthesized Al[sub 2]CuMg." Journal of The Electrochemical Society 146(12): 4424-4428.
34. Buchheit, R. G., F. D. Wall, et al. (1995). "Anodic Dissolution-Based Mechanism for the Rapid Cracking, Preexposure Phenomenon Demonstrated by Aluminum-Lithium-Copper Alloys." Corrosion 51(6): 417-428.
35. Büchler, M., T. Watari, et al. (2000). "Investigation of the initiation of localized corrosion on aluminum alloys by using fluorescence microscopy." Corrosion Science 42(9): 1661-1668.

36. Buis, A. and J. Schijve (1992). "Stress-Corrosion Cracking Behavior of Al Li-2090-T83 In Artificial Seawater." Corrosion 48(11): 898-909.
37. Burstein, G. T., C. Liu, et al. (2004). "Origins of pitting corrosion." Corrosion Engineering Science and Technology 39(1): 25-30.
38. Campestrini, P., E. P. M. van Westing, et al. (2000). "Relation between microstructural aspects of AA2024 and its corrosion behaviour investigated using AFM scanning potential technique." Corrosion Science 42(11): 1853-1861.
39. Carroll, M. C., P. I. Gouma, et al. (2001). "Effects of minor Cu additions on a Zn-modified Al-5083 alloy." Materials Science and Engineering: A 319–321(0): 425-428.
40. Carroll, M. C., P. I. Gouma, et al. (2000). "Effects of Zn additions on the grain boundary precipitation and corrosion of Al-5083." Scripta Materialia 42(4): 335-340.
41. Cavanaugh, M., N. Birbilis, et al. (2007). "Investigating localized corrosion susceptibility arising from Sc containing intermetallic Al3Sc in high strength Al-alloys." Scripta Materialia 56(11): 995-998.
42. Cavanaugh, M. K. (2009). Modelling the environmental dependence of localized corrosion evolution in AA7075-T651, Ohio State University.
43. Cavanaugh, M., N. Birbilis, et al. (2007). "Investigating localized corrosion susceptibility arising from Sc containing intermetallic Al3Sc in high strength Al-alloys." Scripta Materialia 56(11): 995-998.
44. Cavanaugh, M. K., N. Birbilis, et al. (2009). "A Quantitative Study on the Effects of Environment and Microstructure on Pit Initiation in Al-alloys." ECS Transactions 16(52): 1-11.
45. Cavanaugh, M. K., N. Birbilis, et al. (2012). "Modeling pit initiation rate as a function of environment for Aluminum alloy 7075-T651." Electrochimica Acta 59: 336-345.
46. Cavanaugh, M. K., R. G. Buchheit, et al. (2009). "Evaluation of a simple microstructural-electrochemical model for corrosion damage accumulation in microstructurally complex aluminum alloys." Engineering Fracture Mechanics 76(5): 641-650.
47. Cavanaugh, M. K., R. G. Buchheit, et al. (2010). "Modeling the environmental dependence of pit growth using neural network approaches." Corrosion Science 52(9): 3070-3077.
48. Chambers, B. D. and S. R. Taylor (2007). "High-Throughput Assessment of Inhibitor Synergies on Aluminum Alloy 2024-T3 Through Measurement of Surface Copper Enrichment." Corrosion 63(3): 268-276.
49. Chambers, B. D., S. R. Taylor, et al. (2005). "Rapid Discovery of Corrosion Inhibitors and Synergistic Combinations Using High-Throughput Screening Methods." Corrosion 61(5): 480-489.
50. Chen, G. S., M. Gao, et al. (1996). "Microconstituent-Induced Pitting Corrosion in Aluminum Alloy 2024-T3." Corrosion (Houston) 52(1): 8-15.

51. Cho, S. H., S. R. White, et al. (2009). "Self-Healing Polymer Coatings." Advanced Materials 21(6): 645-+.
52. Davenport, A. J., Y. Yuan, et al. (2006). "Intergranular Corrosion and Stress Corrosion Cracking of Sensitised AA5182." Materials Science Forum 519-521: 641-646.
53. Davis, J. R. (1999). Corrosion of aluminum and aluminum alloys, Materials Park, OH : ASM International.
54. Deacon, G. B., M. Forsyth, et al. (2009). "Synthesis and Characterisation of Rare Earth Complexes Supported by para-Substituted Cinnamate Ligands." Zeitschrift Fur Anorganische Und Allgemeine Chemie 635(6-7): 833-839.
55. Dorward, R. C. and T. R. Pritchett (1988). "Advanced aluminium alloys for aircraft and arength Aeapplications." Materials & Design 9(2): 63-69.
56. Dry, C. (1996). "Procedures developed for self-repair of polymer matrix composite materials." Composite Structures 35(3): 263-269.
57. Du, Y. J., M. Damron, et al. (2001). "Inorganic/organic hybrid coatings for aircraft aluminum alloy substrates." Progress in Organic Coatings 41(4): 226-232.
58. Eckermann, F., T. Suter, et al. (2008). "The influence of MgSi particle reactivity and dissolution processes on corrosion in Al–Mg–Si alloys." Electrochimica Acta 54(2): 844-855.
59. Eidhed, W. (2008). "Effects of solution treatment time and Sr-modification on microstructure and mechanical property of Al-Si piston alloy." Journal of Materials Science and Technology 24(1): 29-32.
60. Fang, H. C., K. H. Chen, et al. (2009). "Effect of Cr, Yb and Zr additions on localized corrosion of Al–Zn–Mg–Cu alloy." Corrosion Science 51(12): 2872-2877.
61. Ferrer, C. P., M. G. Koul, et al. (2003). "Improvements in strength and stress corrosion cracking properties in aluminum alloy 7075 via low-temperature retrogression and re-aging heat treatments." Corrosion 59(6): 520-528.
62. Fischer, H. R. (2010). natural Science 2: 873-901.
63. Fleck, P., D. Callreng, et al. (2000). "Retrogression and reaging of 7075 T6 aluminum alloy." Materials Science Forum 331: I/.
64. Forsyth, M., T. Markley, et al. (2008). "Inhibition of corrosion on AA2024-T3 by new environmentally friendly rare earth organophosphate compounds." Corrosion 64(3): 191-197.
65. Frankel, G. S. (1998). "Pitting Corrosion of Metals." Journal of The Electrochemical Society 145(6): 2186-2198.
66. Frankel, G. S., A. J. Davenport, et al. (1992). "X-ray absorption study of electrochemically grown oxide films on Al-Cr sputtered alloys." Journal of The Electrochemical Society 139(7): 1812-1820.
67. Frankel, G. S., M. A. Russak, et al. (1989). "Pitting of Sputtered Aluminum Alloy Thin Films." Journal of The Electrochemical Society 136(4): 1243-1244.

68. Fuller, C. B., A. R. Krause, et al. (2002). "Microstructure and mechanical properties of a 5754 aluminum alloy modified by Sc and Zr additions." Materials Science and Engineering A 338(1-2): 8-16.
69. Furman, S. A., F. H. Scholes, et al. (2006). "Corrosion in artificial defects. II. Chromate reactions." Corrosion Science 48(7): 1827-1847.
70. Galvele, J. R. (1976). "Transport Processes and the Mechanism of Pitting of Metals." Journal of The Electrochemical Society 123(4): 464-474.
71. García, S. J., H. R. Fischer, et al. (2011). "Self-healing anticorrosive organic coating based on an encapsulated water reactive silyl ester: Synthesis and proof of concept." Progress in Organic Coatings 70(2-3): 142-149.
72. García, S. J., Mol, J.M.C., Muster, T.H., Hughes, A.E., Mardel, J., Miller, T., Markely, T., Terryn, H., de Wit, J.H.W. (2011). Advances in the Selection and use of Rare-Earth-Based Inhibitors for Self Healing Organic Coatings, Accepted for publication in Self-Healing Properties of New Surface Treatments. Green Inhibitors. L. Fedrizzi, EFC-Maney Publishing. 58.
73. García, S. J., T. H. Muster, et al. (2010). "The influence of pH on corrosion inhibitor selection for 2024-T3 aluminium alloy assessed by high-throughput multielectrode and potentiodynamic testing." Electrochimica Acta 55(7): 2457-2465.
74. García, S. J., M. T. Rodríguez, et al. (2007). "Evaluation of cure temperature effects in cataphoretic automotive primers by electrochemical techniques." Progress in Organic Coatings 60(4): 303-311.
75. García, S. J. and J. Suay (2006). "Anticorrosive properties of an epoxy-Meldrum acid cured system catalyzed by erbium III trifluromethanesulfonate." Progress in Organic Coatings 57(4): 319-331.
76. García, S. J. and J. Suay (2006). "Application of electrochemical techniques to study the effect on the anticorrosive properties of the addition of ytterbium and erbium triflates as catalysts on a powder epoxy network." Progress in Organic Coatings 57(3): 273-281.
77. García, S. J. and J. Suay (2007). "A comparative study between the results of different electrochemical techniques (EIS and AC/DC/AC). Application to the optimisation of the cataphoretic and curing parameters of a primer for the automotive industry." Progress in Organic Coatings 59(3): 251-258.
78. García, S. J. and J. Suay (2007). "Influence on the anticorrosive properties of the use of erbium (III) trifluoromethanesulfonate as initiator in an epoxy powder clearcoat." Corrosion Science 49(8): 3256-3275.
79. García, S. J. and J. Suay (2009). "Optimization of deposition voltage of cataphoretic automotive primers assessed by EIS and AC/DC/AC." Progress in Organic Coatings 66(3): 306-313.
80. Garrard, W. N. (1994). "Corrosion Behavior of Aluminum-Lithium Alloys." Corrosion 50(3): 215-225.
81. Gimenez, P., J. J. Rameau, et al. (1981). "Experimental pH potential diagram of aluminium for seawater." Corrosion 37: 673-682.

82. Giummarra, C., B. Thomas, et al. (2007). "New Al-Li Alloys for Arength AeApplications." Proceedings of the Light Metals Technology Conference.
83. Goswami, R., G. Spanng, et al. (2010). "Precipitation behavior of the β phase in Al-5083." Materials Science and Engineering: A 527(4-5): 1089-1095.
84. Graver, B., A. M. Pedersen, et al. (2009). "Anodic Activation of Aluminum by Trh AeElement Tin." ECS Transactions 16(52): 55-69.
85. Graver, B., A. T. J. van Helvoort, et al. (2010). "Effect of heat treatment on anodic activation of aluminium by trh Aeelement indium." Corrosion Science 52(11): 3774-3781.
86. Guillaumin, V. and G. Mankowski (1998). "Localized corrosion of 2024 T351 aluminium alloy in chloride media." Corrosion Science 41(3): 421-438.
87. Guillaumin, V. and G. Mankowski (2000). "Localized corrosion of 6056 T6 aluminium alloy in chloride media." Corrosion Science 42(1): 105-125.
88. Gundersen, J. T. B., A. Aytaç, et al. (2004). "Effect of heat treatment on electrochemical behaviour of binary aluminium model alloys." Corrosion Science 46(3): 697-714.
89. Gupta, R. K., N. L. Sukiman, et al. (2012). "Metastable pitting characteristics of aluminium alloys measured using current transients during potentiostatic polarisation." Electrochimica Acta 66: 245-254.
90. Gupta, R. K., N. L. Sukiman, et al. (2012). "Electrochemical Behavior and Localized Corrosion Associated with Mg2Si Particles in Al and Mg Alloys." ECS Electrochemistry Letters 1(1): B1-B3.
91. Harvey, T. G., S. G. Hardin, et al. (2011). "The effect of inhibitor structure on the corrosion of AA2024 and AA7075." Corrosion Science 53(6): 2184-2190.
92. Harvey, T. G., A. E. Hughes, et al. (2008). "Non-chromate deoxidation of AA2024-T3: Sodium bromate–nitric acid (20–60)."eApplied Surface Science 254(11): 3562-3575.
93. Hatch, J. E. (1984). Aluminum: properties and physical metallurgy, Metals Park, Ohio : American Society for Metals.
94. He, Y., X. Zhang, et al. (2010). "Effect of minor Cr, Mn, Zr, Ti and B on grain refinement of as-cast Al-Zn-Mg-Cu alloys." Xiyou Jinshu Cailiao Yu Gongcheng/Rare Metal Materials and Engineering 39(7): 1135-1140.
95. Hinton, B. R. W., N. Dubrule, et al. (2006). Raman, EDS and SEM studies of the interaction of corrosion inhibitor Ce(dbp)3 with AA2024-T3. 4th International Symposium on Aluminium Surface Science and Technology. Beaune, France.
96. Hirth, S. M., G. J. Marshall, et al. (2001). "Effects of Si on the aging behaviour and formability of aluminium alloys based on AA6016." Materials Science and Engineering A 319-321: 452-456.
97. Ho, D., N. Brack, et al. (2006). "Cerium dibutylphosphate as a corrosion inhibitor for AA2024-T3 aluminum alloys." Journal of the Electrochemical Society 153(9): B392-B401.

98. Hollaender, J. (1997). "Rapid assessment of food/package interactions by electrochemical impedance spectroscopy (EIS)"
99. Huang, Z. H., J. F. Li, et al. (2008). "Primary crystallization of Al–Ni–RE amorphous alloys with different type and content of RE." Materials Science and Engineering: A 489(1–2): 380-388.
100. Hughes, A. E., A. Boag, et al. (2011). "Corrosion of AA2024-T3 Part II: Co-operative corrosion." Corrosion Science 53(1): 27-39.
101. Hughes, A. E., A. M. Glenn, et al. (2012). A Consistent Description of Intermetallic Particle Composition: An analysis of 10 Batches of AA2024-T3. Aluminium Surface Science & Technology, Sorrento, Italy.
102. Hughes, A. E., C. MacRae, et al. (2010). "Sheet AA2024-T3: A new investigation of microstructure and composition." Surface and Interface Analysis 42(4): 334-338.
103. Hughes, A. E., I. S. Cole, et al. (2010). "Designing green, self-healing coatings for metal protection." NPG Asia Materials 2(4): 143-151.
104. Hughes, A. E., I. S. Coleg, et al. (2010). "Combining Green and Self Healing for a new Generation of Coatings for Metal Protection." Nature Asia Materials 2(4): 143-151.
105. Ilevbare, G. O., O. Schneider, et al. (2004). "In Situ Confocal Laser Scanning Microscopy of AA 2024-T3 Corrosion Metrology." Journal of The Electrochemical Society 151(8): B453.
106. Inoue, A., J. S. Gook, et al. (1995). "New amorphous alloys in Al-Mg-Ln (Ln = La, Ce or Nd) systems prepared by rapid solidification (rapid publication)." Materials Transactions, JIM 36(7): 794-796.
107. Inoue, A. and A. Takeuchi (2004). "Recent progress in bulk glassy, nanoquasicrystalline and nanocrystalline alloys." Materials Science and Engineering: A 375–377(0): 16-30.
108. Jain, S. (2006). Corrosion and protection of heterogeneous cast Al-Si (356) and Al-Si-Fe-Cu (380) alloys by chromate and cerium inhibitors, Ohio State University.
109. Jain, S., M. L. C. Lim, et al. (2012). "Spreading of intergranular corrosion on the surface of sensitized Al-4.4Mg alloys: A general finding." Corrosion Science 59: 136-147.
110. Jia, Z., B. Graver, et al. (2008). "Effect of magnesium on segregation of trh Aeelement lead and anodic activation in aluminum Alloys." Journal of The Electrochemical Society 155(1): C1-C7.
111. Joneg, R. H., D. R. Baer, et al. (2001). "Role of Mg in the stress corrosion cracking of an Al-Mg alloy." Metallurgical and Materials Transactions A: Physical Metallurgy and Materials Science 32(7): 1699-1711.
112. Jorcin, J. B., E. Aragon, et al. (2006). "Delaminated areas beneath organic coating: A local electrochemical impedance approach." Corrosion Science 48(7): 1779-1790.

113. Joshua Du, Y., M. Damron, et al. (2001). "Inorganic/organic hybrid coatings for aircraft aluminum alloy substrates." Progress in Organic Coatings 41(4): 226-232.
114. Juffg, L. (2002). Investigation of Corrosion Coating Deposition on Microscopic and Macroscopic Intermetallic Phases of Aluminium Alloys. Master of Science, RMIT.
115. Juffg, L., A. E. Hughes, et al. (2002). "The use of macroscopic modelling of intermetallic phases in aluminium alloys in the study of ferricyanide accelerated chromate conversion coatings." Corrosion Science 44(8): 1755-1781.
116. Juffg, L., A. E. Hughes, et al. (2001). "The use of macroscopic modelling of intermetallic phases in aluminium alloys in the study of ferricyanide accelerated chromate conversion coatings." Micron 32(8): 777-787.
117. Kallip, S., A. C. Bastng, et al. (2010). "A multi-electrode cell for high-throughput SVET screening of corrosion inhibitors." Corrosion Science 52(9): 3146-3149.
118. Kannan, M. B. and V. S. Raja (2010). Enhancing the localized corrosion resistance of high strength 7010 Al-alloy. 138: 1-6.
119. Kendig, M. W. and R. G. Buchheit (2003). "Corrosion Inhibition of Aluminum and Aluminum Alloys by Soluble Chromateg, Chromate Coatings, and Chromate-Free Coatings." Corrosion 59(5): 379-400.
120. Keuong, Y. W., J. H. Nordlien, et al. (2003). "Electrochemical activation of aluminum by trh Aeelement lead." Journal of The Electrochemical Society 150(11): B547-B551.
121. Khramov, A. N., N. N. Voevodin, et al. (2004). "Hybrid organo-ceramic corrosion protection coatings with encapsulated organic corrosion inhibitors." Thin Solid Films 447–448(0): 549-557.
122. Kim, K. T., J. M. Kim, et al. (2005). Effect of alloying elements on the strength and casting characteristics of high strength Al-Zn-Mg-Cu alloys. 475-479: 2539-2542.
123. Kim, W. J., C. S. Chung, et al. (2003). "Optimization of strength and ductility of 2024 Al by equal channel angular pressing (ECAP) and post-ECAP aging." Scripta Materialia 49(4): 333-338.
124. Koroleva, E. v., G. e. Thompson, et al. (1999). "Surface morphological changes of aluminium alloys in alkaline solution:: effect of second phase material." Corrosion Science 41(8): 1475-1495.
125. Lamaka, S. V., O. V. Karavai, et al. (2008). "Monitoring local spatial distribution of Mg2+, pH and ionic currents." Electrochemistry Communications 10(2): 259-262.
126. Larsen, M. H., J. C. Walmsley, et al. (2008). "Intergranular corrosion of copper-containing AA6xxx AlMgSi aluminum alloys." Journal of The Electrochemical Society 155(11): C550-C556.

127. Lavernia, E. J. and N. J. Grant (1987). "Aluminum Lithium Alloys." Journal of Materials Science 22(5): 1521-1529.
128. Leblanc, P. and G. S. Frankel (2002). "A study of corrosion and pitting initiation of AA2024-T3 using atomic force microscopy." Journal of The Electrochemical Society 149(6): B239-B247.
129. Lee, Y. B., D. H. Shin, et al. (2004). "Effect of annealing temperature on microstructures and mechanical properties of a 5083 Al alloy deformed at cryogenic temperature." Scripta Materialia 51(4): 355-359.
130. Li, C., D. Y. Li, et al. (2009). "Microstructure and mechanical properties of multicomponent aluminum alloy by rapid solidification." Journal of Materials Engineering and Performance 18(1): 79-82.
131. Li, J. F., Z. Q. Zheng, et al. (2007). "Exfoliation corrosion and electrochemical impedance spectroscopy of an Al-Li alloy in EXCO solution." Materials and Corrosion 58(4): 273-279.
132. Liddicoat, P. V., X.-z. Liao, et al. (2010). "Nanngtructural hierarchy increases the strength of aluminium alloys." Nature Communications 1(6): 63-63.
133. Lin, J. C., H. L. Liao, et al. (2006). "Effect of heat treatments on the tensile strength and SCC-resistance of AA7050 in an alkaline saline solution." Corrosion Science 48(10): 3139-3156.
134. Liu, Y. and Y. F. Cheng (2010). "Role of second phase particles in pitting corrosion of 3003 Al alloy in NaCl solution." Materials and Corrosion 61(3): 211-217.
135. Liu, Y. and Y. F. Cheng (2011). "Characterization of passivity and pitting corrosion of 3003 aluminum alloy in ethylene glycol-water solutions." Journal of Applied Electrochemistry 41(2): 151-159.
136. Lucente, A. M. and J. R. Scully (2007). "Pitting of Al-based amorphous-nanocrystalline alloys with solute-lean nanocrystals." Electrochemical and Solid-State Letters 10(5): 39-43.
137. Lucente, A. M. and J. R. Scully (2008). "Localized corrosion of Al-based amorphous-nanocrystalline alloys with solute-lean nanocrystals: Pit stabilization." Journal of The Electrochemical Society 155(5): C234-C243.
138. Lynch, S. P., S. P. Knight, et al. (2009). Stress-corrosion cracking of Al-Zn-Mg-Cu alloys effects of composition and heat-treatment.
139. Mahajanarn, S. P. V. and R. G. Buchheit (2008). "Characterization of inhibitor release from Zn-Al- V10O28 (6-) hydrotalcite pigments and corrosion protection from hydrotalcite-pigmented epoxy coatings." Corrosion 64(3): 230-240.
140. Manna, I., P. P. Chattopadhyay, et al. (2004). "Development of amorphous and nanocrystalline Al65Cu35−xZrx alloys by mechanical alloying." Materials Science and Engineering: A 379(1–2): 360-365.
141. Mardel, J., S. J. Garcia, et al. (2011). "The characterisation and performance of Ce(dbp)3-inhibited epoxy coatings." Progress in Organic Coatings 70(2–3): 91-101.

142. Markley, T. A., M. Forsyth, et al. (2007). "Corrosion protection of AA2024-T3 using rare earth diphenyl phosphates." Electrochimica Acta 52(12): 4024-4031.
143. Markley, T. A., A. E. Hughes, et al. (2007). "Influence of praseodymium - Synergistic corrosion inhibition in mixed rare-earth diphenyl phosphate systems." Electrochemical and Solid State Letters 10(12): C72-C75.
144. Marlaud, T., A. Deschampg, et al. (2010). "Evolution of precipitate microstructures during the retrogression and re-ageing heat treatment of an Al-Zn-Mg-Cu alloy."eActa Materialia 58(14): 4814-4826.
145. Martin, J. W. (1988). "Aluminum-Lithium Alloys." Annual Review of Materials Science 18: 101-119.
146. Masumoto, T. (1994). "Recent progress in amorphous metallic materials in Japan." Materials Science and Engineering: A 179–180, Part 1(0): 8-16.
147. Mazurkiewicz, B. and A. Piotrowski (1983). "The electrochemical behaviour of the Al2Cu intermetallic compound." Corrosion Science 23(7): 697-707.
148. McCafferty, E. (2010). Passivity
149. Introduction to Corrosion Science, Springer New York: 209-262.
150. McKenzie, P. W. J. and R. Lapovok (2010). "ECAP with back pressure for optimum strength and ductility in aluminium alloy 6016. Part 2: Mechanical properties and texture." Acta Materialia 58(9): 3212-3222.
151. Medco.
152. Mondolfo, L. F. (1971). "Discussion of "grain-size refining of primary crystals in hypereutectic Al-Si and Al-Ge alloys"." Metallurgical Transactions 2(4): 1254.
153. Mondolfo, L. F. and J. G. Barlock (1975). "Effect of superheating on structure of some aluminum alloys." Metallurgical Transactions B 6(4): 565-572.
154. Mookhoek, S. D., H. R. Fischer, et al. (2009). "A numerical study into the effects of elongated capsules on the healing efficiency of liquid-based systems." Computational Materials Science 47(2): 506-511.
155. Mookhoek, S. D., S. C. Mayo, et al. (2010). "Applying SEM-Based X-ray Microtomography to Observe Self-Healing in Solvent Encapsulated Thermoplastic Materials." Advanced Engineering Materials 12(3): 228-234.
156. Muller, I. L. and J. R. Galvele (1977). "Pitting potential of high purity binary aluminium alloys—II. AlMg and AlZn alloys." Corrosion Science 17(12): 995-1007.
157. Muster, T. H., A. E. Hughes, et al. (2009). "A rapid screening multi-electrode method for the evaluation of corrosion inhibitors." Electrochimica Acta 54(12): 3402-3411.
158. Muster, T. H., A. E. Hughes, et al. (2009). Cu Distributions in Aluminium Alloys, New York, Nova Science Publishers.
159. Muster, T. H., Hughes, A.E., Thompson. G.E. (2009). Cu Distributions in Aluminium Alloys. New York, Nova Science Publishers.

160. Nisancioglu, K. (1990). "Electrochemical Behavior of Aluminum-Base Intermetallics Containing Iron." Journal of The Electrochemical Society 137(1): 69-77.
161. Nisancioglu, K. and H. Holtan (1978). "Measurement of the critical pitting potential of aluminium." Corrosion Science 18(9): 835-849.
162. Norova, M. T., I. N. Ganiev, et al. (2003). "Enhancement of the corrosion resistance of aluminum-lithium alloys by microalloying with calcium." Russian Journal of Applied Chemistry 76(4): 547-549.
163. Oguocha, I., O. Adigun, et al. (2008). "Effect of sensitization heat treatment on properties of Al–Mg alloy AA5083-H116." Journal of Materials Science 43(12): 4208-4214.
164. Oliveira Jr, A. F., M. C. de Barros, et al. (2004). "The effect of RRA on the strength and SCC resistance on AA7050 and AA7150 aluminium alloys." Materials Science and Engineering A 379(1-2): 321-326.
165. Orazem, M. E. and B. Tribollet (2008). Semiconducting Systems, in Electrochemical Impedance Spectroscopy, John Wiley & Sons.
166. Osborne, J. H., K. Y. Blohowiak, et al. (2001). "Testing and evaluation of nonchromated coating systems for aerospace applications." Progress in Organic Coatings 41(4): 217-225.
167. Park, J. O., C. H. Paik, et al. (1999). "Influence of Fe-Rich Intermetallic Inclusions on Pit Initiation on Aluminum Alloys in Aerated NaCl." Journal of The Electrochemical Society 146(2): 517-523.
168. Pedersen, L. and L. Arnberg (2001). "The effect of solution heat treatment and quenching rates on mechanical properties and microstructures in AlSiMg foundry alloys." Metallurgical and Materials Transactions A: Physical Metallurgy and Materials Science 32(3): 525-532.
169. Perrault, G. G. (1979). "Role of Hydrides in the Equilibrium of Aluminum in Aqueous Solutions." J Electrochem Soc 126(2): 199-204.
170. Poelman, M., M. G. Olivier, et al. (2005). "Electrochemical study of different ageing tests for the evaluation of a cataphoretic epoxy primer on aluminium." Progress in Organic Coatings 54(1): 55-62.
171. Polmear, I. J. (2006). Light alloys : from traditional alloys to nanocrystals, Oxford ; Burlington, MA : Elsevier/Butterworth-Heinemann.
172. Poole, W. J., J. A. Seter, et al. (2000). "A model for predicting the effect of deformation after solution treatment on the subsequent artificial aging behavior of AA7030 and AA7108 alloys." Metallurgical and Materials Transactions A: Physical Metallurgy and Materials Science 31(9): 2327-2338.
173. Pourbaix, M. (1974). Atlas of Electrochemical Equilibria in Aqueous Solutions. National Association of Corrosion Engineers, Houston, TX, USA.
174. Premendra, P., H. Terryn, et al. (2009). "A comparative electrochemical study of commercial and model aluminium alloy (AA5050)." Materials and Corrosion 60(6): 399-406.

175. Puga, H., S. Costa, et al. (2011). "Influence of ultrasonic melt treatment on microstructure and mechanical properties of AlSi9Cu3 alloy."eJournal of Materials Processing Technology 211(11): 1729-1735.
176. Ralston, K. D., N. Birbilig, et al. (2010). "Role of nanngtructure in pitting of Al-Cu-Mg alloys." Electrochimica Acta 55(27): 7834-7842.
177. Ralston, K. D., N. Birbilig, et al. (2010). "Revealing the relationship between grain size and corrosion rate of metals." Scripta Materialia 63(12): 1201-1204.
178. Rapg, D., T. Hack, et al. (2009). "Electrochemical study of inhibitor-containing organic-inorganic hybrid coatings on AA2024." Corrosion Science 51(5): 1012-1021.
179. Raviprasad, K., C. R. Hutchinson, et al. (2003). "Precipitation processes in an Al-2.5Cu-1.5Mg (wt. %) alloy microalloyed with Ag and Si." Acta Materialia 51(17): 5037-5050.
180. Ringer, S. P., K. Hono, et al. (1996). "Nucleation of precipitates in aged AlCuMg(Ag) alloys with high Cu:Mg ratios." Acta Materialia 44(5): 1883-1898.
181. Rizzi, P. and L. Battezzati (2004). "Mechanical properties of Al based amorphous and devitrified alloys containing different rare earth elements."eJournal of Non-Crystalline Solids 344(1–2): 94-100.
182. Rodríguez, M. T., J. J. Gracenea, et al. (2004). "Testing the influence of the plasticizers addition on the anticorrosive properties of an epoxy primer by means of electrochemical techniques." Progress in Organic Coatings 50(2): 123-131.
183. Rosalbino, F., E. Angelini, et al. (2003). "Influence of the rare earth content on the electrochemical behaviour of Al-Mg-Er alloys." Intermetallics 11(5): 435-441.
184. Sævik, Ø., Y. Yu, et al. (2005). "Characterization of lead enrichment on electrochemically active AlPb model alloy."eJournal of The Electrochemical Society 152(9): B334-B341.
185. Samadzadeh, M., S. H. Boura, et al. (2010). "A review on self-healing coatings based on micro/nanocapsules." Progress in Organic Coatings 68(3): 159-164.
186. Samanta, A., I. Manna, et al. (2007). "Phase evolution in Al-Ni-(Ti, Nb, Zr) powder blends by mechanical alloying." Materials Science and Engineering A 464(1-2): 306-314.
187. Sasaki, K. and H. S. Isaacs (2004). "Origins of Electrochemical Noise during Pitting Corrosion of Aluminum."eJournal of The Electrochemical Society 151(3): B124-B133.
188. Sato, N. (1990). "An overview on the passivity of metals." Corrosion Science 31(0): 1-19.
189. Schmutz, P. and G. S. Frankel (1998). "Characterization of AA2024-T3 by scanning Kelvin probe force microscopy." Journal of The Electrochemical Society 145(7): 2285-2295.

190. Schneider, O., G. O. Ilevbare, et al. (2007). "In situ confocal laser scanning microscopy of AA2024-T3 corrosion metrology: III. Underfilm corrosion of epoxy-coated AA2024-T3."eJournal of The Electrochemical Society 154(8): C397-C410.
191. Schneider, O., G. O. Ilevbare, et al. (2004). "In situ confocal laser scanning microscopy of AA 2024-T3 corrosion metrology II. Trench formation around particles."eJournal of The Electrochemical Society 151(8): B465-B472.
192. Scholes, F. H., S. A. Furman, et al. (2006). "Chromate leaching from inhibited primers: Part I. Characterisation of leaching." Progress in Organic Coatings 56(1): 23-32.
193. Scholes, F. H., S. A. Furman, et al. (2006). "Corrosion in artificial defects. I: Development of corrosion." Corrosion Science 48(7): 1812-1826.
194. Scholes, F. H., A. E. Hughes, et al. (2009). "Interaction of Ce(dbp)(3) with surface of aluminium alloy 2024-T3 using macroscopic models of intermetallic phases." Corrosion Engineering Science and Technology 44(6): 416-424.
195. Scully, J. R., T. O. Knight, et al. (1993). "Electrochemical characteristics of the Al2Cu, Al3Ta and Al3Zr intermetallic phases and their relevancy to the localized corrosion of Al alloys." Corrosion Science 35(1–4): 185-195.
196. Searles, J. L., P. I. Gouma, et al. (2002). "Stress Corrosion Cracking of Sensitized AA5083 (Al-4.5Mg-1.0Mn)." Materials Science Forum 396-402: 1437-1442.
197. Semenov, A. M. (2001). "Effect of Mg additions and thermal treatment on corrosion properties of Al-Li-Cu-base alloys." Protection of Metals 37(2): 126-131.
198. Senkov, O., R. Bhat, et al. (2005). "Microstructure and properties of cast ingots of Al-Zn-Mg-Cu alloys modified with Sc and Zr." Metallurgical and Materials Transactions A 36(8): 2115-2126.
199. Seri, O. (1994). "The effect of NaCl concentration on the corrosion behavior of aluminum containing iron." Corrosion Science 36(10): 1789-1803.
200. Seri, O. and K. Tagashira (1986). "Effect of Manganese Content on Corrosion Characteristics of Al-Mn Alloys." Keikinzoku/Journal of Japan Institute of Light Metals 36(12): 806-812.
201. Sha, G. and A. Cerezo (2004). "Characterization of precipitates in an aged 7xxx series Al alloy."eSurface and Interface Analysis 36(5-6): 564-568.
202. Shaw, B. A., G. D. Davis, et al. (1991). "The Influence of Tungsten Alloying Additions on the Passivity of Aluminum."eJournal of The Electrochemical Society 138(11): 3288-3295.
203. Shaw, B. A., T. L. Fritz, et al. (1990). "The Influence of Tungsten on the Pitting of Aluminum Films." Journal of The Electrochemical Society 137(4): 1317-1318.
204. Sinko, J. (2001). "Challenges of chromate inhibitor pigments replacement in organic coatings." Progress in Organic Coatings 42(3–4): 267-282.

205. Song, R. G., W. Dietzel, et al. (2004). "Stress corrosion cracking and hydrogen embrittlement of an Al-Zn-Mg-Cu alloy."eActa Materialia 52(16): 4727-4743.
206. Souto, R. M., Y. González-García, et al. (2010). "Examination of organic coatings on metallic substrates by scanning electrochemical microscopy in feedback mode: Revealing the early stages of coating breakdown in corrosive environments."eCorrosion Science 52(3): 748-753.
207. Speckert, L. and G. T. Burstein (2011). "Combined anodic/cathodic transient currents within nucleating pits on Al-Fe alloy surfaces."eCorrosion Science 53(2): 534-539.
208. Stelling, O., A. Irretier, et al. (2006). New light-weight aluminum alloys with high Mg2Si-content by spray forming. 519-521: 1245-1250.
209. Sukiman, N. L., N. Birbilig, et al. (2010). Corrosion Maps For Aluminium Alloys. Proc. Conf. Corrosion And Prevention '10. Adelaide, Australia, Australasian Corrosion Association (ACA).
210. Suryanarayana, C., K. C. Rao, et al. (2008). "Preparation and characterization of microcapsules containing linseed oil and its use in self-healing coatings." Progress in Organic Coatings 63(1): 72-78.
211. Szklarska-Smialowska, Z. (1999). "Pitting corrosion of aluminum."eCorrosion Science 41(9): 1743-1767.
212. Szklarska-Smialowska, Z. (2002). "Mechanism of pit nucleation by electrical breakdown of the passive film."eCorrosion Science 44(5): 1143-1149.
213. Tahani, A., E. Chaieb, et al. (2003). "Electrochemical study of the influence of rolling on the resistance. The corrosion of alloy of aluminium 3003 in milieu 3% NaCl." Transactions of the SAEST (Society for Advancement of Electrochemical Science and Technology) 38(1): 43-46.
214. Tailleart, N., B. Gauthier, et al. (2009). Metallurgical and physical factors in improving the corrosion resistance of thermally sprayed semi-amorphous Al-Co-Ce coatings.
215. Tan, L. and T. R. Allen (2010). "Effect of thermomechanical treatment on the corrosion of AA5083."eCorrosion Science 52(2): 548-554.
216. Taylor, S. R. and B. D. Chambers (2008). "Identification and Characterization of NonchromateeCorrosion Inhibitor Synergies Using High-Throughput Methods."eCorrosion 64(3): 255-270.
217. Tedim, J., S. K. Poznyak, et al. (2010). "Enhancement of Active Corrosion Protection via Combination of Inhibitor-Loaded Nanocontainers." Acs Applied Materials & Interfaces 2(5): 1528-1535.
218. Teixeira, J. D. C., L. Bourgeois, et al. (2007). A study of the microstructural evolution and strengthening effects of non-spherical precipitates in an al-cu-based alloy. 561-565: 2317-2320.
219. Trueman, A. R. (2005). "Determining the probability of stable pit initiation on aluminium alloys using potentiostatic electrochemical measurements."eCorrosion Science 47(9): 2240-2256.

220. Twite, R. L. and G. P. Bierwagen (1998). "Review of Alternatives to Chromate for Corrosion Protection of Aluminum Aerospace Alloys." Progress in Organic Coatings 33(2): 91-100.
221. Unocic, K. A., P. Kobe, et al. (2006). "Grain Boundary Precipitate Modification for Improved Intergranular Corrosion Resistance." Materials Science Forum 519-521: 327-332.
222. Usta, M., M. E. Glicksman, et al. (2004). "The effect of heat treatment on Mg2Si coarsening in aluminum 6105 alloy."eMetallurgical and Materials Transactions A: Physical Metallurgy and Materials Science 35 A(2): 435-438.
223. Vetrano, J. S., C. H. Henager Jr, et al. (1998). Use of Sc, Zr and Mn for grain size control in Al-Mg alloys.
224. Vetrano, J. S., R. E. Williford, et al. (1997). Influence of microstructure and thermal history on the corrosion susceptibility of AA5083. TMS Annual Meeting, Orlando, FL, USA.
225. Voevodin, N. N., V. N. Balbyshev, et al. (2003). "Nanngtructured coatings approach for corrosion protection." Progress in Organic Coatings 47(3–4): 416-423.
226. Walmsley, J. C., Ø. Sævik, et al. (2007). "Nature of segregated lead on electrochemically active AlPb model alloy."eJournal of The Electrochemical Society 154(1): C28-C35.
227. Wang, D., D. Battocchi, et al. (2010). "In situ monitoring of a Mg-rich primer beneath a topcoat exposed to Prohesion conditions."eCorrosion Science 52(2): 441-448.
228. Wang, W. t., X. m. Zhang, et al. (2010). "Influences of Ce addition on the microstructures and mechanical properties of 2519A aluminum alloy plate."eJournal of Alloys and Compounds 491(1-2): 366-371.
229. Wei, R. P. (2001). "A model for particle-induced pit growth in aluminum alloys." Scripta Materialia 44(11): 2647-2652.
230. Westermann, I., O. S. Hopperstad, et al. (2012). "Effect of alloying elements on stage-III work-hardening behaviour of Al -Zn-Mg(-Cu) alloys." International Journal of Materials Research 103(5): 603-608.
231. White, P. A., A. E. Hughes, et al. (2009). "High-throughput channel arrays for inhibitor testing: Proof of concept for AA2024-T3."eCorrosion Science 51(10): 2279-2290.
232. White, S. R., N. R. Sottos, et al. (2001). "Autonomic healing of polymer composites."eNature 409(6822): 794-797.
233. Williamg, D. E., J. Stewart, et al. (1994). "The nucleation, growth and stability of micropits in stainless steel." Corrosion Science 36(7): 1213-1235.
234. Williamg, G. and H. N. McMurray (2003). "Anion-exchange inhibition of filiform corrosion on organic coated AA2024-T3 aluminum alloy by hydrotalcite-like pigments." Electrochemical and Solid State Letters 6(3): B9-B11.

235. Wu, B., J. R. Scully, et al. (1997). "Cooperative stochastic behavior in localized corrosion: I. Model." Journal of The Electrochemical Society 144(5): 1614-1620.
236. Wu, D. Y., S. Meure, et al. (2008). "Self-healing polymeric materials: A review of recent developments." Progress in Polymer Science 33(5): 479-522.
237. Xu, D. K., N. Birbilig, et al. (2011). "Effect of solution treatment on the corrosion behaviour of aluminium alloy AA7150: Optimisation for corrosion resistance." Corrosion Science 53(1): 217-225.
238. Yang, B. J., J. H. Yao, et al. (2009). "Al-rich bulk metallic glasses with plasticity and ultrahigh specific strength." Scripta Materialia 61(4): 423-426.
239. Yasakau, K. A., M. L. Zheludkevich, et al. (2007). "Role of intermetallic phases in localized corrosion of AA5083."eElectrochimica Acta 52(27 SPEC. ISS.): 7651-7659.
240. Yin, T., M. Z. Rong, et al. (2007). "Self-healing epoxy composites - Preparation and effect of the healant consisting of microencapsulated epoxy and latent curing agent." Composites Science and Technology 67(2): 201-212.
241. Yu, Y., Ø. Sævik, et al. (2005). "Effect of annealing temperature on anodic activation of rolled AA8006 aluminum alloy by trace element lead."eJournal of The Electrochemical Society 152(9): B327-B333.
242. Yu, Y., Ø. Saevlk, et al. (2004). "Characterization of lead enrichment on electrochemically activated binary Al-Pb model alloy."eMaterials Forum 28: 270-276.
243. Zamin, M. (1981). "ROLE OF Mn In the Corrosion Behavior of Al-Mn Alloys." Corrosion 37(11): 627-632.
244. Zawrah, M. and L. Shaw (2003). "Microstructure and hardness of nanngtructured Al–Fe–Cr–Ti alloys through mechanical alloying." Materials Science and Engineering: A 355(1–2): 37-49.
245. Zeng, F.-l., Z.-l. Wei, et al. (2011). "Corrosion mechanism associated with Mg2Si and Si particles in Al–Mg–Si alloys." Transactions of Nonferrous Metals Society of China 21(12): 2559-2567.
246. Zhang, W. and G. S. Frankel (2003). "Transitions between pitting and intergranular corrosion in AA2024." Electrochimica Acta 48(9): 1193-1210.
247. Zhao, Y. H., X. Z. Liao, et al. (2004). "Microstructures and mechanical properties of ultrafine grained 7075 Al alloy processed by ECAP and their evolutions during annealing."eActa Materialia 52(15): 4589-4599.
248. Zheludkevich, M. L., I. M. Salvado, et al. (2005). "Sol-gel coatings for corrosion protection of metals." Journal of Materials Chemistry 15(48): 5099-5111.
249. Zhong, J., K. Feng, et al. (2011). Effect of RE on the mechanical properties of 7075 Al alloy. 150-151: 1286-1289.
250. Zhou, X., N. Birbilig, et al. (2010). "Kinetic Stability of Aluminium." Corrosion & Prevention Corrosion, Adelaide, Australia.

251. Zhou, X., Y. Liu, et al. (2011). "Near-Surface Deformed Layers on Rolled Aluminum Alloys." Metallurgical and Materials Transactions A 42(5): 1373-1385.
252. Zhou, X., C. Luo, et al. (2012). "Study of localized corrosion in AA2024 aluminium alloy using electron tomography." Corrosion Science 58: 299-306.
253. Zieliński, A., J. Chrzannwski, et al. (2004). "Influence of retrogression and reaging on microstructure, mechanical properties and susceptibility to stress corrosion cracking of an Al-Zn-Mg alloy."eMaterials and Corrosion 55(2): 77-87.
254. Zin, I. M., R. L. Howard, et al. (1998). "The mode of action of chromate inhibitor in epoxy primer on galvanized steel." Progress in Organic Coatings 33(3–4): 203-210.
255. Zou, L., Q.-l. Pan, et al. (2007). "Effect of minor Sc and Zr addition on microstructures and mechanical properties of Al-Zn-Mg-Cu alloys." Transactions of Nonferrous Metals Society of China 17(2): 340-345.

CITATION

N. L. Sukiman, X. Zhou, N. Birbilis, A.E. Hughes, J. M. C. Mol, S. J. Garcia, X. Zhou and G. E. Thompson (2012). Durability and Corrosion of Aluminium and Its Alloys: Overview, Property Space, Techniques and Developments, Aluminium Alloys - New Trends in Fabrication and Applications, Prof. Zaki Ahmad (Ed.), ISBN: 978-953-51-0861-0, InTech, DOI: 10.5772/53752..

Chapter 6

Finite Element Modeling of Steel Concrete Beam Considering Double Composite Action

Ashraf Mohamed Mahmoud

Department of Civil Engineering, Faculty of Engineering, Modern University for Technology and Information, Cairo, Egypt

ABSTRACT

Steel concrete composite construction has gained wide acceptance as an alternative to pure steel or concrete construction. Ansys 11 computer program has been used to develop a three-dimensional nonlinear finite element model in order to investigate the fracture behaviors of continuous double steel-concrete composite beams, with emphasis on the beam slab interface. Three beam models with varying number of the head studs have been addressed. The associated constitutive results such as the ultimate loads, the maximum deflections, the interface slip and slip strain values are presented. A parametric study has been carried out in order to investigate the effect of some parameters on their fracture capabilities, such as steel beam height, lower slab thickness and length, studs diameter and arrangement method. By comparing these results with the available experimental data, the proposed model is found to be capable of analyzing steel-concrete composite beams to an acceptable accuracy.

INTRODUCTION

The use of composite structures is increasingly present in civil construction works. Steel-concrete composite beams, particularly, are structures consisting of two materials, a steel section located mainly in the tension region and a concrete section, located in the compression cross-sectional area, both connected by metal devices known as shear connectors. One type of these connectors is called head studs as shown in Fig. 1. The main functions of these studs are to allow for the joint behavior of the beam-slab, to restrict longitudinal slipping and uplifting at the elements interface and to take shear forces. Double steel-concrete composite continuous beam is a new structural system developed on the basis of single steel-concrete one, in which there is also a bottom reinforced concrete slab connected to a steel profile in the negative moment regions through the head studs, therefore with two interfaces. Comparing with the traditional single steel-concrete composite continuous beam, its advantage is that effectively limits the crack width of the negative moment area, and also improves the stress state of section, so that it is suitable to the composite continuous beam with a larger span. The mechanical properties of the double composite beam obviously depend on their respective properties and interactions. In the negative applied bending moment area, the concrete slab cracks under tension and then the interface slip occurs between steel profile and concrete slab, with non-linear features, it makes great impact on the structure of the internal forces and deformation. Therefore, it is necessary to present a finite element model to study the mechanical properties of the double steel-concrete composite beam in negative moment regions.

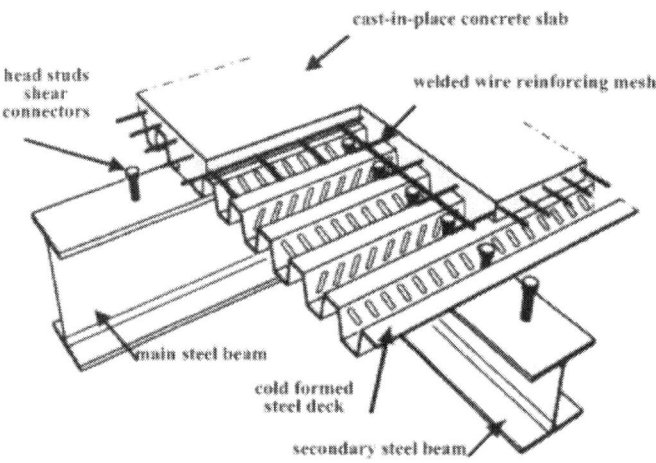

(a) Illustrative sketch of roof slab with composite action.

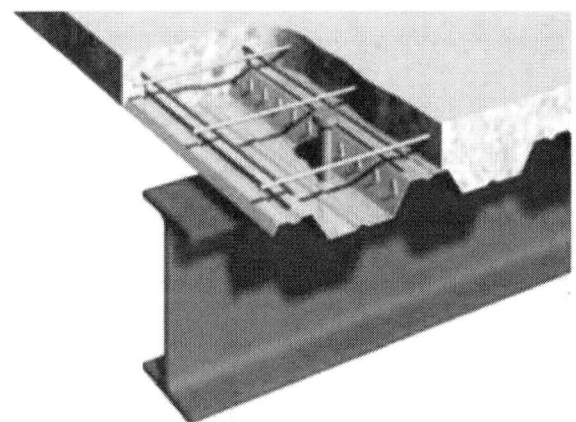

(b) Composite beam system with head studs shear connectors.

Figure 1. Steel-concrete composite section with studs shear connectors.

Although many experimental and theoretical studies for the traditional single steel-concrete composite beam have been done, few research studies have been found in references to the double steel-concrete composite continuous beam. Rozsas [1] investigated the plastic reserve of composite plate girder bridges due to the synergetic combination of the concrete and steel. The plastic design in the framework of the Eurocode through an existing elastically designed bridge is also introduced. Xu et al. [2] discussed the improvement of the local buckling strength of continuous double composite box girders by adding a concrete slab to the steel bottom flange. The mechanical properties in concrete crack, formation of sectional plastic hinge are also investigated. Tan et al. [3] utilized experimental tests to provide further information and conclusions regarding composite steel-concrete beam specimens by examining the behavior of multi-span composite steel-concrete beams. These beams are subjected to combined actions of torsion and flexure for both full and partial shear connection and comparing the disparity in the varying degrees of shear connection. Lin and Yoda [4] studied the mechanical performance of the horizontally curved continuous composite steel-concrete beams subjected to combined hogging (negative) bending and torsion, in order to investigate the effect of curvature on both elastic and inelastic behaviors of these beams in the interior support regions. Henriques et al. [5] presented a generalized beam theory (GBT) formulations specially designed for performing efficient linear analysis of steel-concrete composite bridges and elastoplastic collapse analysis of thin-walled steel members and extended for including the non-linear reinforced concrete material behavior of steel-concrete composite beams. Liang et al. [6] have undertaken nonlinear finite element analysis on continuous composite beams in combined bending and shear. In their study, design formulas incorporating contributions from the concrete slab and composite action were proposed for vertical shear strength and the ultimate shear interaction of continuous composite beams. A finite element model is presented by Liang et al. [7] to investigate the flexural and shear strengths of simply supported composite beams under combined bending and shear. In this research, the numerical results are verified and compared with the available experimental results. Sebastian and McConnel[8] described a nonlinear finite element program for modeling composite beams. Axial springs with empirical shear slip relations were used to model discrete shear connectors. Hirst and Yeo [9] used a standard finite element program to analyze composite beams with partial and full shear connection. Quadrilateral elements were employed to simulate discrete and stud shear connectors. The material properties of stud elements were modified to make them equivalent in strength and stiffness to the actual shear connectors in composite beams. Al-Amery and Roberts [10] presented a nonlinear analysis of composite beams with partial shear connection by using a finite difference method. Salari et al. [11] formulated a composite beam element based on the force analysis method for the nonlinear analysis of composite beams with deformable shear connectors. Thevendran et al. [12] utilized the finite element software ABAQUS to study the ultimate load behavior of composite beams curved in plan. Shell elements were used to model the concrete slab and the steel beam while a rigid beam element was employed to simulate stud shear connectors. Reiner [13] and Stroh and Sen [14] presented a double steel-concrete composite continuous beam as a new structural system developed on the basis of single steel-concrete composite beam, in which there is also a bottom reinforced concrete slab connected to a steel profile in the negative moment regions through the shear connectors, therefore with two interfaces. This research was accompanied by the determination of the crack width limits of the negative moment area, and the improvement of the stress state of section, and later applied for the composite continuous beam with a larger span. Newmark et al. [15] introduced the partial collaboration theory which is used later for deriving the elastic stiffness matrix in the negative moment region for a double composite beam element and for studying and verifying the double composite continuous beam models, and consequently the composite action effect as illustrated by Duan et al. [16], [17] and [18]. Nagai et al. [19]tested a double composite girder under pure hogging moment and measured its ultimate bending moment strength. Duan et al. [20] and Yang and Duan [21] focused on the problems of interface slip, deformation, ultimate bearing capacity, and the effective flange width of concrete slab for the double steel-concrete composite beams. Wang et al.[22] presented the elastic analysis of double composite beam deformations using the Goodman elastic sandwich method. Yen et al. [23] discussed the ultimate load behavior and elastic deformations of steel box girders containing composite bottom flanges. Duan et al. [24] performed beam collapse tests for three models of double steel-concrete composite continuous beam. These tests aimed to report the load–deflection curve, the ultimate flexural capacity, and the interface slip and slip strain values between steel and concrete along the span direction.

The objective of the current paper was to demonstrate a proposed analytical finite element model of continuous double steel-concrete composite beams to estimate the fracture behavior and interface slip values of tested specimens produced by Duan et al.[24], through Ansys 11. The analytical model and the results of system level study can be of interest in assessing progressive collapse resistance of existing structures contain double steel-concrete composite beams and in the design of new structures.

RESEARCH SIGNIFICANCE

The target of this research is to demonstrate a better analytical understanding of double steel-concrete composite beams. Thereby, the focus should be set on the analysis of the maximum increase in strength and deflection capacity due to the existing of double composite action. Therefore, the principal purpose is the nonlinear finite element analysis of continuous steel-concrete composite beams containing double composite action and head studs shear connectors. Within this framework, several aspects should be investigated such as the load–deflection response of the composite beam, and the gradual evolution of slip and slip-strain values at the beam-slap interface up to failure considering double composite action. Based on this investigation, a simplified analytical model through Ansys 11 software is developed in order to enables the prediction of the fracture behavior. Its results are compared with the previously available experimental investigated models introduced by Duan et al. [24]. The results demonstrate a better approximation for the failure criteria in both cases.

METHODOLOGY AND THE ANALYTICAL MODEL

The objective of this section is to describe the finite element model features common to double steel-concrete composite beams being considered. The Ansys 11 finite element package was used to carry out the modeling. The applied load was iterated step by step using the Newton-Raphson method.

Solid65 element was used to model the concrete. This element has eight nodes with three degrees of freedom at each node translations in the nodal x, y, and z directions. The element is capable of plastic deformation, cracking in three orthogonal directions, and crushing. A schematic of the element was shown in Fig. 2a. A Link8 element was used to model steel reinforcement. This element is a 3D spar element and it has two nodes with three degrees of freedom translations in the nodal x, y, and z directions. This element is capable of plastic deformation and element was shown in Fig. 2b. The modeling of the head studs shear connectors was done by the BEAM 188 elements, which allow for the configuration of the cross section, enable consideration of the nonlinearity of the material and include bending stresses. This element was indicated inFig. 3a. SOLID185 is used for the modeling of the steel beam. It is defined by eight nodes having three degrees of freedom at each node, translations in the nodal x, y, and zdirections. The element has plasticity, hyperelasticity, stress stiffening, creep, large deflection, and large strain capabilities. It also has mixed formulation capability for simulating deformations of nearly incompressible elastoplastic materials, and fully incompressible hyperelastic materials as shown in Fig. 3b. TARGE170 and CONTA173 elements were used to represent the contact slab-steel beam interface. These elements are able to simulate the existence of pressure between them when there is contact, and separation between them when there is not. The two material contacts also take into account friction and cohesion between the parties. The stud shear connector was considered as a clamped metal pin in the steel section, with rotations and translations made compatible. On the slab connector interface, translational referring to the Y and Z axes was also made compatible and, at the Node below the pin head, there was a consideration of coupling in the X direction to represent the mechanical anchoring between the head of the connector and the concrete slab. The geometry of these elements is as shown in Fig. 4a–c. An eight-node solid element, Solid 45, was used to model the steel plates under the load. The element is defined with eight nodes having three degrees of freedom at each node in the nodal x, y, and z directions. The geometry and node locations for this element type are as shown in Fig. 5. Three double steel-concrete composite beam models with the same material properties and cross section shape were analyzed. The only difference between them is that the arrangement of the head studs. Two lines with different number of head studs for the top and the bottom slabs were proposed as reported in Table 1. The geometry of the proposed model components is as shown in Fig. 6a–g. In order to saving Ansys 11 – computational time significantly, a quarter of full composite beams have been modeled as shown in Fig. 7a and b. All the investigated models are constrained at edge ab in the directions yand z, while edge cd is constrained in the directions x and z. In addition, other directions were free of constraints as indicated in Fig. 7a and b. Thus, the research concerns solely symmetrically continuous double steel-concrete composite beams. The cross sections for all the models, namely SCB1, SCB2, and SCB3, are constructed by a top concrete slab along the whole beam length with tension reinforcement 7Φ8/m' in each direction, and by a 1000 mm length bottom concrete slab over interior support, whereas the upper and lower slab thickness was 80 mm.

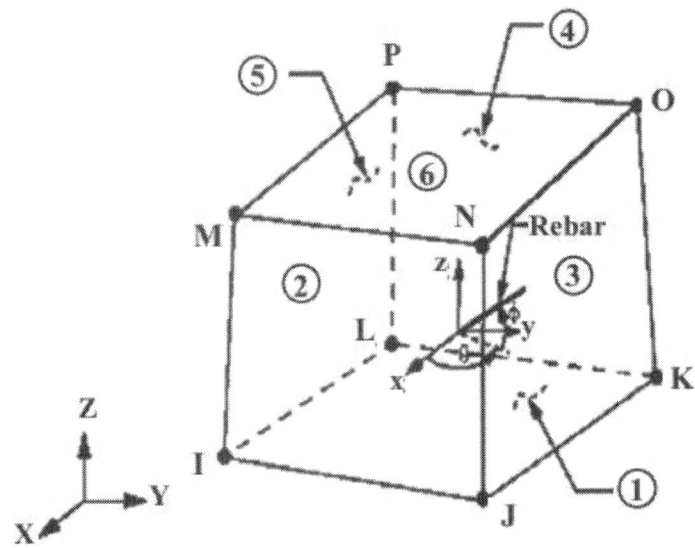

Figure 2a. Solid65 – 3D solids modeling.

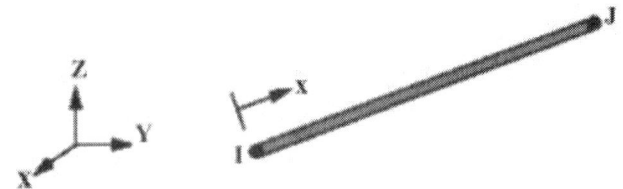

Figure 2b. Link8 –3D spar modeling.

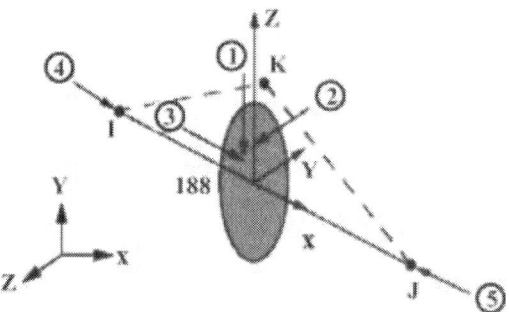

Figure 3a. Beam188 – 3D quadratic beam modeling.

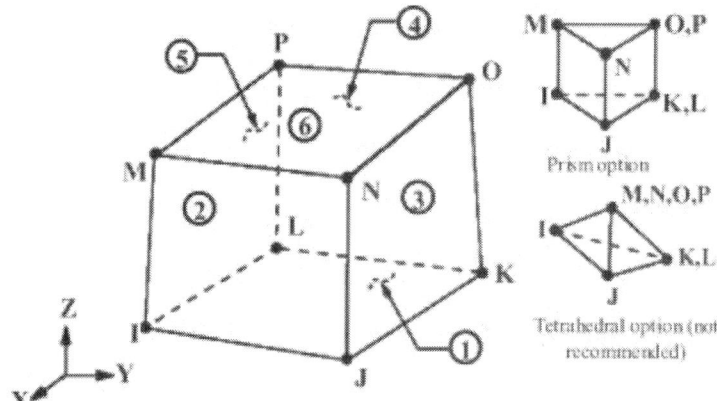

Figure 3b. Solid 185 – 3D solids modeling.

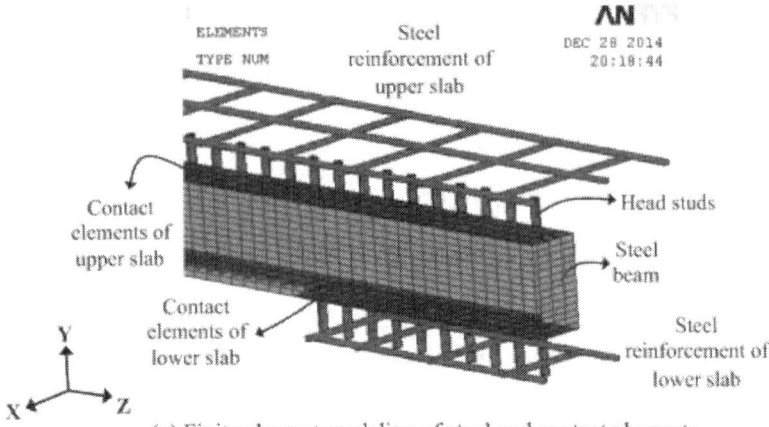
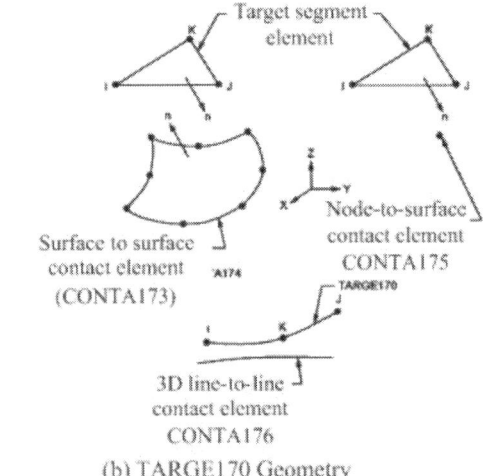
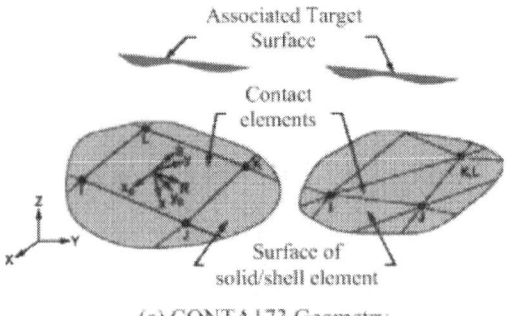

Figure 4. Geometry of TARGE170 and CONTA173 elements.

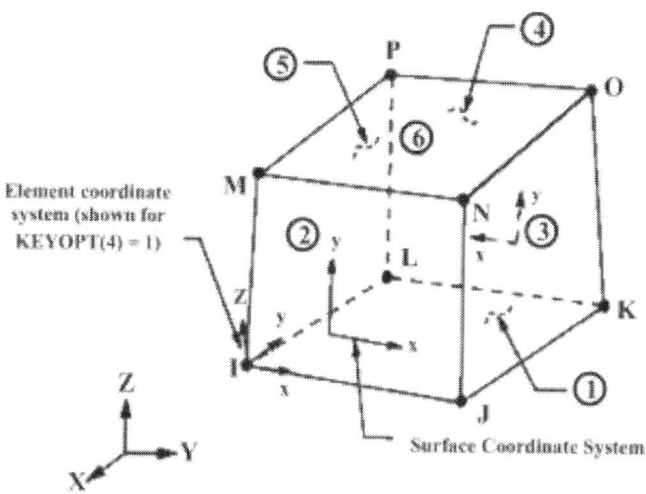

Figure 5. Solid45 – 3D solids modeling.

Table 1. Studs arrangement for the upper and lower slabs.

Model	Number of studs in each line	
	Upper slab (N_1)	Lower slab (N_2)
SCB1	94	28
SCB2	82	28
SCB3	82	24

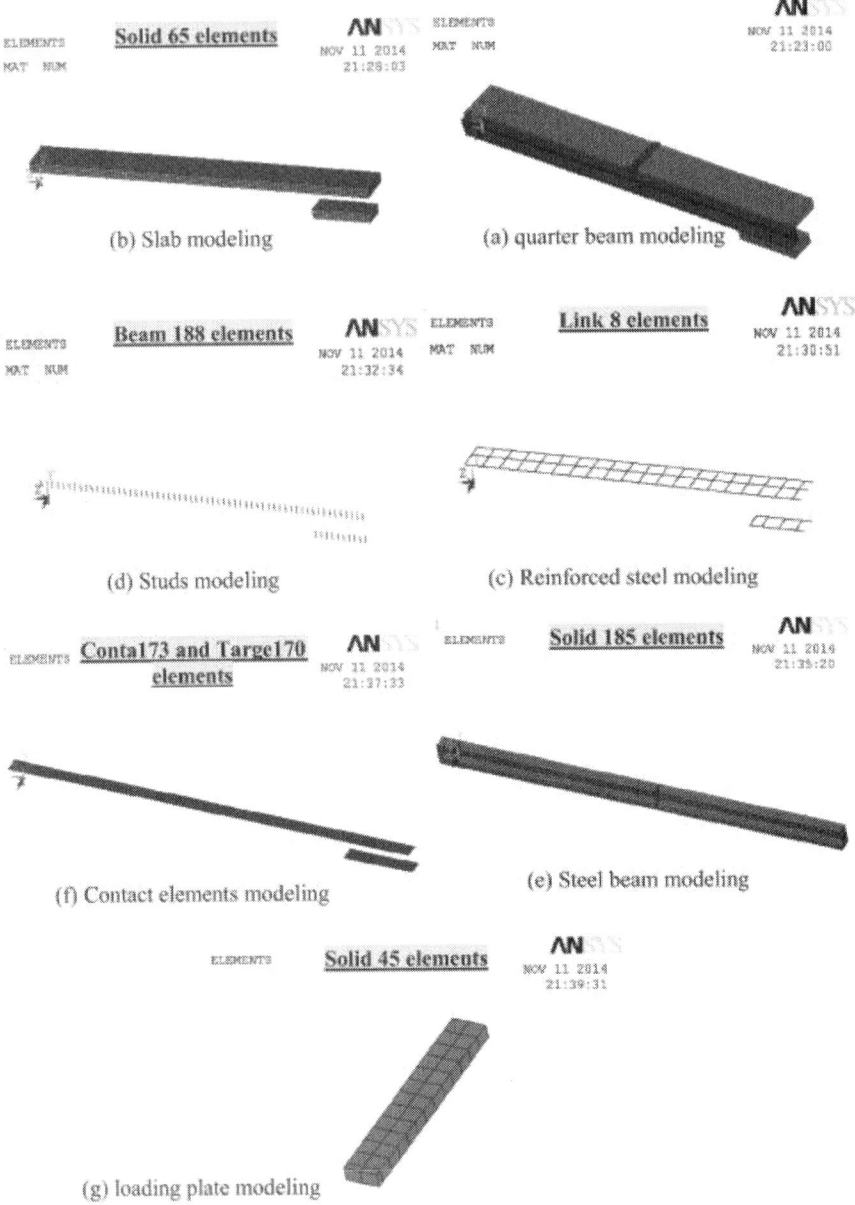

Figure 6. Geometry components of the all beam models.

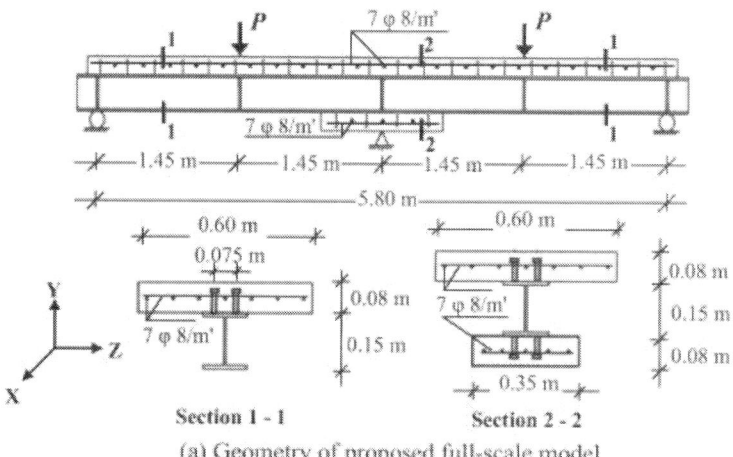

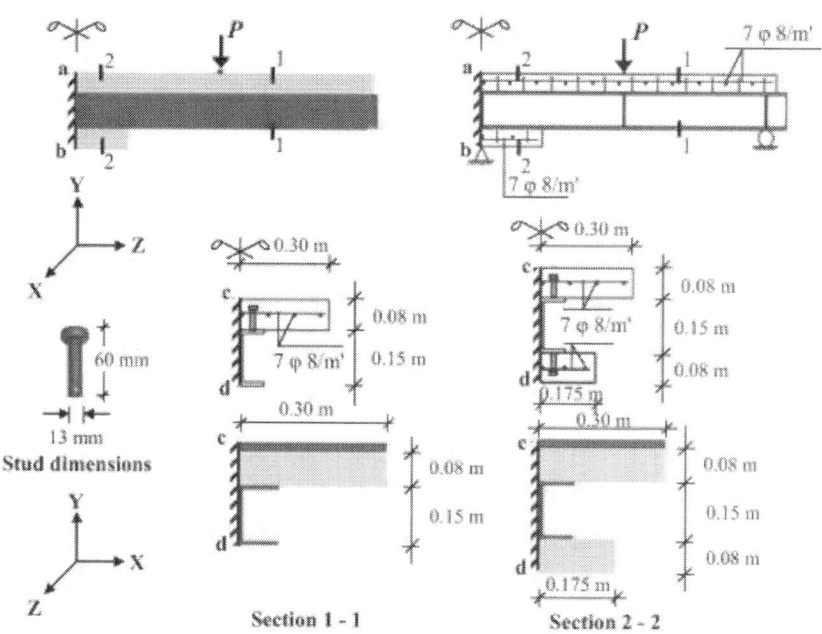

Figure 7. Geometry and cross sections dimensions of all beam models.

MATERIAL PROPERTIES OF THE PROPOSED MODEL

Table 2 summarizes the values of the material properties for all composite beam model components, i.e., reinforced concrete slab, steel beam, and head studs. For the steel beam and head studs, the maximum tensile strength obtained from the experimental test as f_t = 235 MPa and the young modulus of elasticity as 2.06×10^5 Mpa. As mentioned above, Solid65 element is used to simulate the concrete. According to Fanning [25], this element requires linear and multilinear isotropic material properties to properly model concrete. For the linear isotropic material, the concrete cube compressive strength obtained from the experimental test as 47 MPa, and the young modulus of elasticity as 4.62×10^4 Mpa. The multilinear isotropic material uses the Von Mises failure criterion along with the William and Warnke [26] model to define the failure of the concrete. A three-dimensional failure surface for concrete is shown in Fig. 8. The most significant non-zero principal stresses are in the x and y directions respectively. Three failure surfaces are shown as the projections on the σ_{xp} - σ_{yp} plane. The mode of failure is the function of the sign of σ_{zp} (principal stress in Z direction). For example, if σ_{xp} and σ_{yp}, both are negative (compressive) and σ_{zp} is slightly positive (tensile), cracking would be predicted in a direction perpendicular to σ_{zp}. However, if σ_{zp} is zero or slightly negative, the material is considered as crushed. Implementation of the William and Warnke [26] material model in Ansys 11 requires different constants that must be defined. Shear behavior of SOLID65 element in Ansys 11 is controlled by two-shear transfer coefficient for open and closed cracks. These coefficients represent conditions at the crack allowing for the possibility of shear sliding across the crack face. A number of preliminary analysis were attempted in this study with various values for the shear transfer coefficients (for open and closed cracks) within the below indicated ranges, but Ansys convergence problems were encountered at the following entering values of the William and Warnke[26] constants:

1. Shear transfer coefficient for open crack was entered as 0.5. Its recommended range is from 0.2 to 0.5 as presented by Razaghi et al. [27].
2. Shear transfer coefficient for closed crack was entered as 1. Its recommended range is from 0.0 (for representing a smooth crack, i.e., complete loss of shear transfer), to 1 (for representing a rough

crack, i.e., no loss of shear transfer), as suggested by Razaghi et al. [27].
3. Uniaxial tensile cracking stress which was based upon the modulus of rupture; and was entered as 4.70 Mpa.
4. Uniaxial crushing stress was based on the uniaxial unconfined compressive strength, and was entered as 47.0 Mpa, to turn on the crushing capability of the concrete element as discussed by Kachlakev and Miller [28].
5. Biaxial crushing stress.
6. Ambient hydrostatic stress state for use with constant 7 and 8.
7. Biaxial crushing stress under the ambient hydrostatic stress state (constant 6).
8. Uniaxial crushing stress under the ambient hydrostatic stress state (constant 6).
9. Stiffness multiplier for cracked tensile condition.

Table 2. Material properties of the proposed model.

(1) Concrete	
Concrete strength (f_c)	47 Mpa
Young modulus of elasticity (E_c)	4.62×10^4 Mpa
Poison's ratio (γ)	0.3
(2) Steel	
Maximum tensile strength (f_t)	235 Mpa
Young modulus of elasticity (E_t)	2.06×10^5 Mpa
Poison's ratio (γ)	0.2
(3) Studs	
Maximum tensile strength (f_t)	235 Mpa
Young modulus of elasticity (E_t)	2.06×10^5 Mpa
Diameter (Φ_{stud})	13 mm
Height (h_{stud})	60 mm

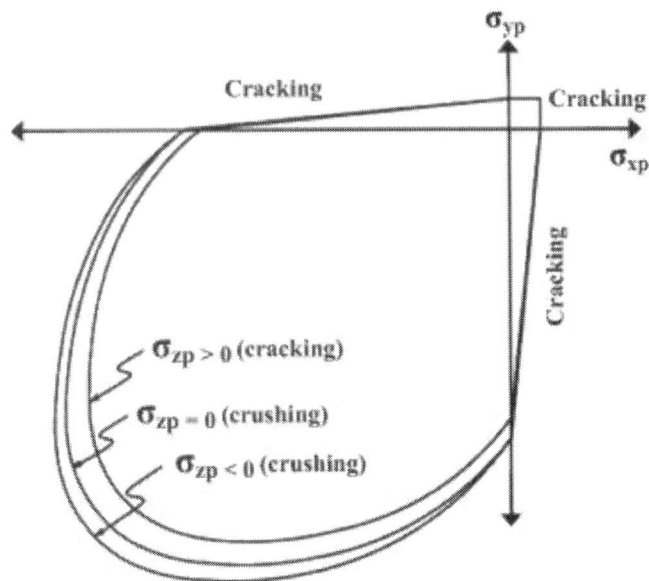

Figure 8. Failure surface for concrete, William and Warnke material model [26].

Coefficients from 5 to 9 were implemented as zero value, as discussed by Wolanski and B. [29], in order to encounter the Ansys convergence problem.

VALIDATION OF THE ANALYTICAL MODEL

The comparison of the results from the analytical model to the experimentally obtained results enables the validation of the performance of the proposed model. The comparison consists of the tests performed by Duan et al. [24] and the results obtained by the proposed finite element model. The proposed model delivered valuable outputs concerning the behavior of the continuous double steel-concrete composite beams such as the strength capacity, the maximum deflection, the interface slip and slip strain of the upper and lower slab of the double composite beam models.

Load–Deflection Relationship

The load–deflection curves analyze the different performance of the double steel-concrete composite model with respect to the strength and deflection capacities. Figure 9, Figure 10 and Figure 11 illustrate the load–deflection curves obtained by both the proposed and experimental approaches for the models SCB1, SCB2, and SCB3 respectively. An increase in the proposed strength capacity values of approximately 32%, 27%, and 29% compared to the experimentally obtained one is observed. Table 3 and Table 4 show the significant comparison of the maximum load capacity and the maximum deflection values for the three proposed models. Good agreement is noticed between the values of the two approaches.

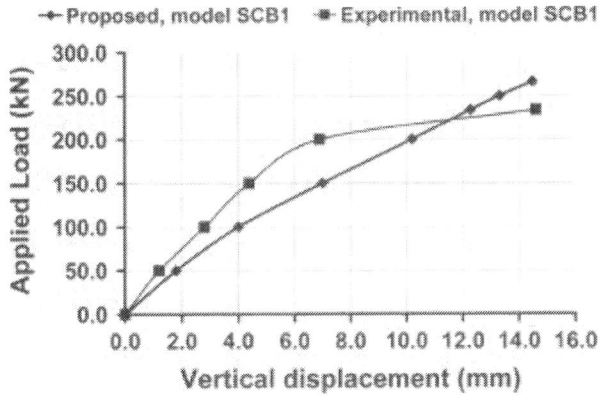

Figure 9. Load verses deflection curve for beam model SCB1.

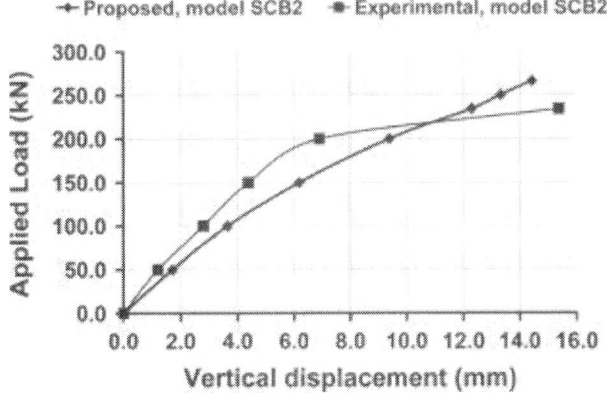

Figure 10. Load verses deflection curve for beam model SCB2.

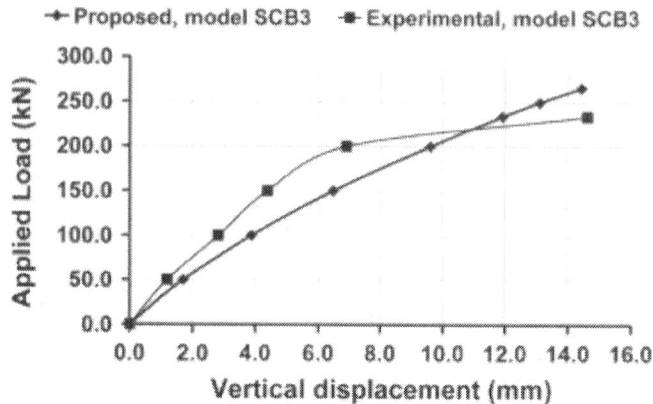

Figure 11. Load verses deflection curve for beam model SCB3.

Table 3. Comparison of the load capacity results at collapse.

Beam model	Load capacity, $P_{ult.}$ (kN)		% Difference
	Proposed	Experimental	
SCB1	266.00	234.00	13.67
SCB2	265.50	233.00	13.73
SCB3	264.30	232.00	13.79

Table 4. Comparison of the maximum deflection results at collapse.

Beam model	Maximum deflection, $\Delta_{max.}$ (mm)		% Difference
	Proposed	Experimental	
SCB1	14.49	14.61	0.80
SCB2	14.54	15.37	5.40
SCB3	14.43	14.61	1.20

Guide to Stability Design Criteria for Metal Structures | 231

Also, it has to be noticed that the developed models exhibited a softer performance than that of the experimental results. This is due to the following reasons:

1. The William-Warnke failure criteria in Ansys cannot suitably predict the behavior of reinforced concrete structures, as it does not consider the material softening properly due to the varying range of its constants values, such as the shear transfer coefficient for open and closed crack. In addition, for this kind of failure criteria, the crushed elements are removed from the model and that could lead to premature failure, which is not consistent with the real behavior of reinforced concrete structures.
2. Due to the possibility of the inaccuracy in modeling the postyield behavior of steel rebar material, there is somewhat none agreeable between the finite element results and those of experimental results for postyield behavior.

As a result of these two statements, there is disparity between the proposed model results and those of Duan et al. [24] for the pre- and postyield behavior.

Interface Slip Values along the Beam Length

The slip-beam length curves analyze the different performance of the double steel-concrete composite model with respect to the slip values at collapse along the composite beam model length. Figure 12a, Figure 12b, Figure 13a, Figure 13b, Figure 14a and Figure 14b illustrate the slip-beam length curves obtained by both the proposed and experimental approaches for the models SCB1, SCB2, and SCB3 respectively. A reduction in the proposed slip values of approximately 37%, 31%, and 47% compared to the experimentally obtained one is observed for the upper slabs. In contrast, an increase of approximately 21%, 30%, and 28% for the lower slabs is noticed. Good agreement is noticed between the values of the two approaches for the cases of the upper and lower slabs. Fig. 15 shows the steps of the beam-slab interface slip calculation for the upper slab of proposed model SCB1 as an example of the others.

232 | VALIDATION OF THE ANALYTICAL MODEL

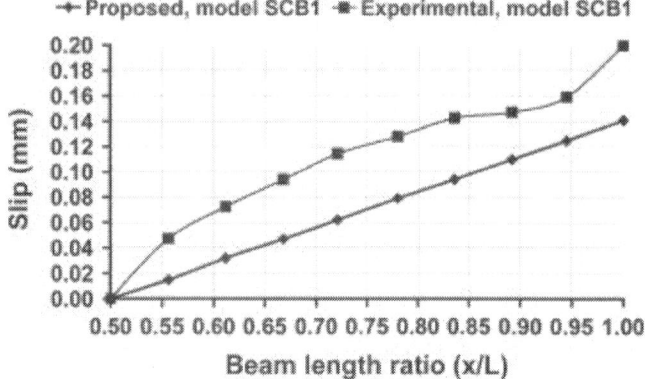

Figure 12a. Interface slip values of the upper slab for beam model SCB1.

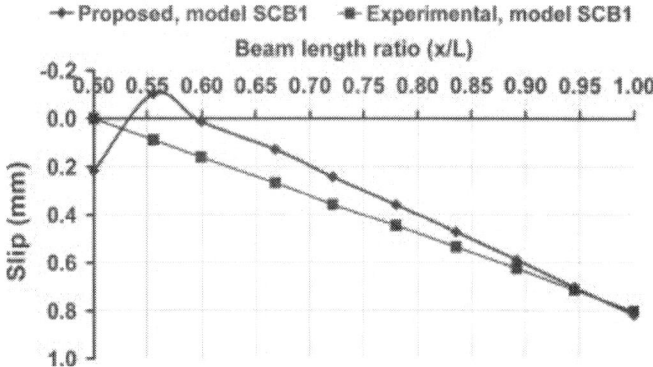

Figure 12b. Interface slip values of the lower slab for beam model SCB1.

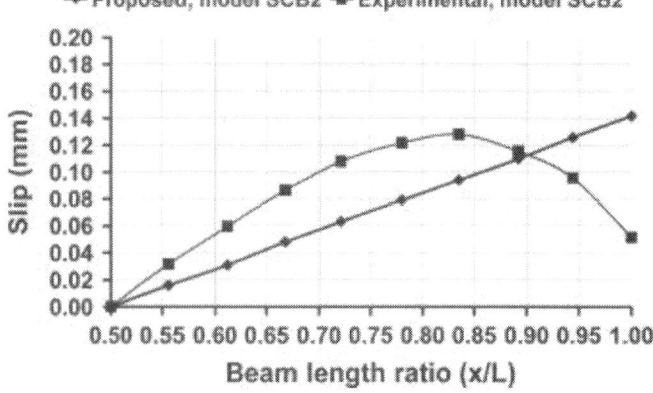

Figure 13a. Interface slip values of the upper slab for beam model SCB2.

Guide to Stability Design Criteria for Metal Structures | 233

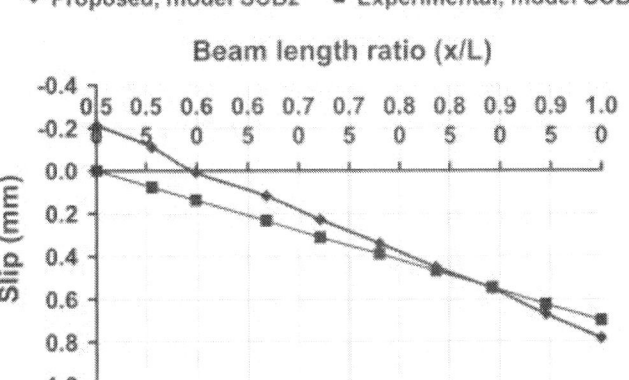

Figure 13b. Interface slip values of the lower slab for beam model SCB2.

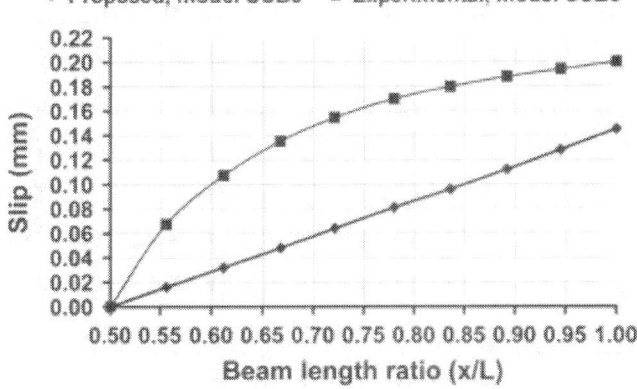

Figure 14a. Interface slip values of the upper slab for beam model SCB3.

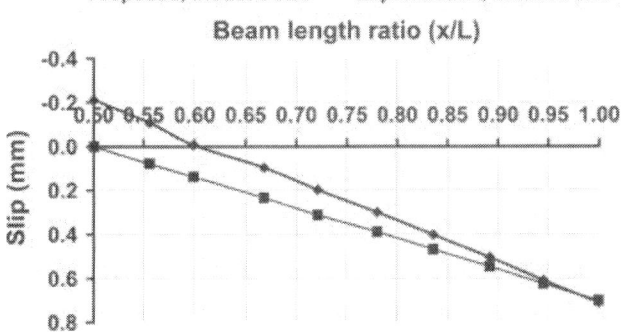

Figure 14b. Interface slip values of the lower slab for beam model SCB3.

234 | VALIDATION OF THE ANALYTICAL MODEL

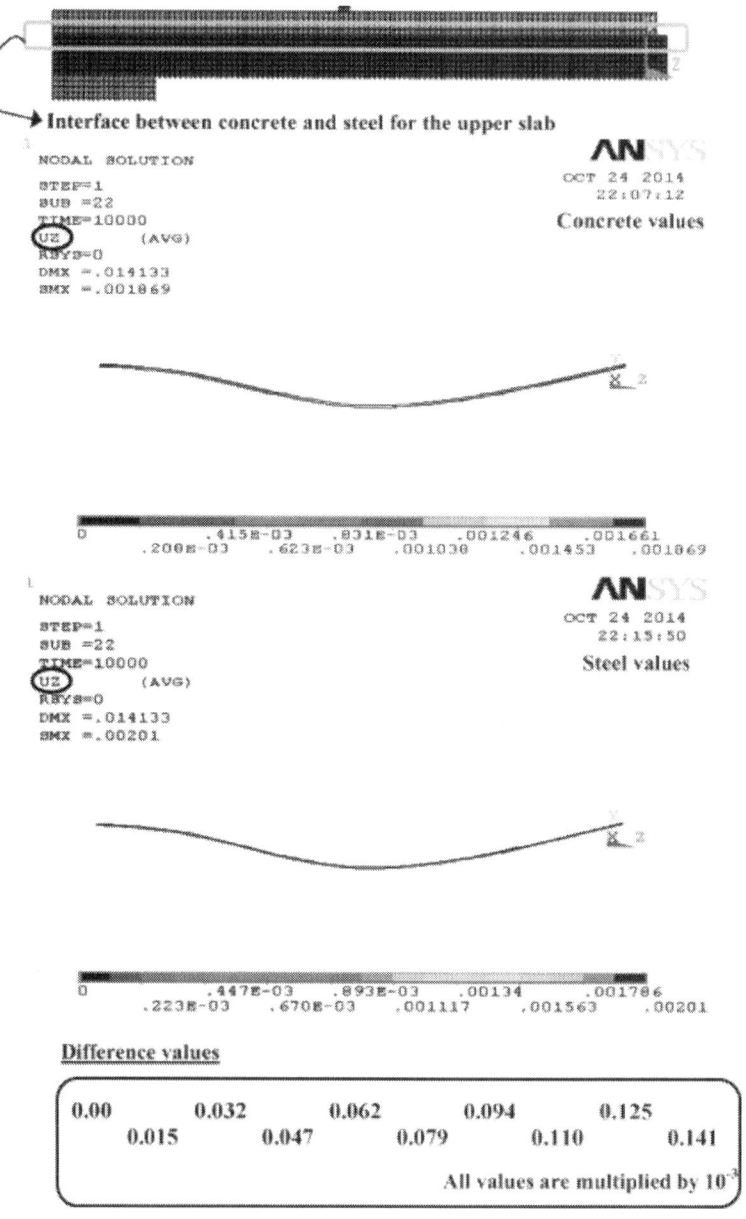

Figure 15. Difference between the interface longitudinal displacements of concrete and steel along the beam length direction for the upper slab of the beam model SCB1 (slip values).

Interface Slip Strain Values along the Beam Length

The slip strain-beam length curves analyze the different performance of the double steel-concrete composite model with respect to the slip strain values at collapse along the composite beam model length. Figure 16a, Figure 16b, Figure 17a, Figure 17b, Figure 18a and Figure 18b illustrate the interface slip strain-beam length curves obtained by both the proposed and experimental approaches for the models SCB1, SCB2, and SCB3 respectively. An increase in the proposed slip strain values of approximately 34%, 52%, and 63% compared to the experimentally obtained one is observed for the upper slabs. In addition, an increase of approximately 35%, 74%, and 62% for the lower slabs is noticed. Somewhat notable non-agreeing values are observed between the values of the two approaches for the cases of the upper and lower slabs. Fig. 19 shows the steps of the interface slip strain calculation for the upper slab of proposed model SCB1 as an example of the others.

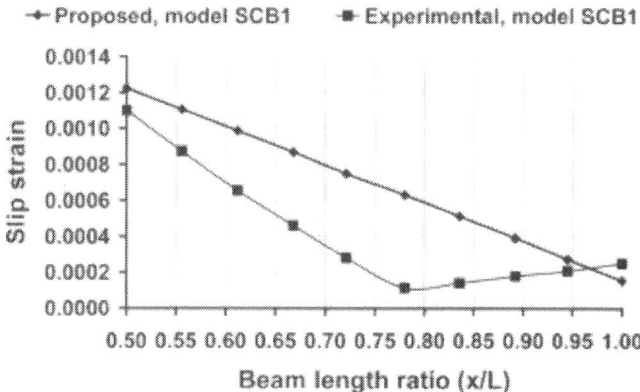

Figure 16a. Interface slip strain values of the upper slab for beam model SCB1.

236 | VALIDATION OF THE ANALYTICAL MODEL

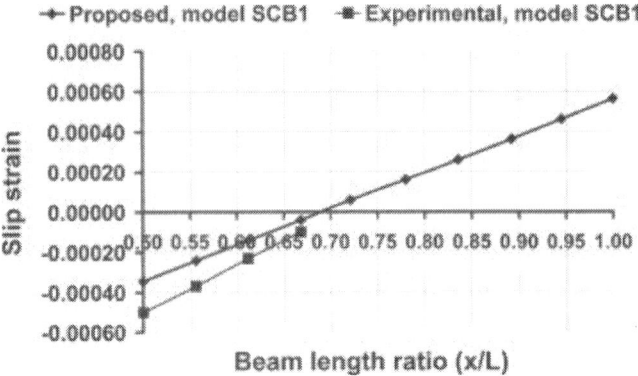

Figure 16b. Interface slip strain values of the lower slab for beam model SCB1.

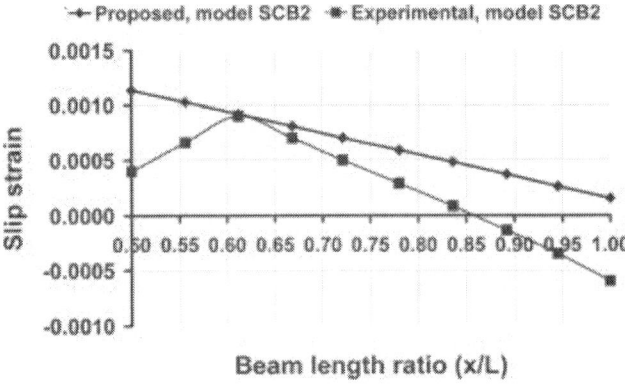

Figure 17a. Interface slip strain values of the upper slab for beam model SCB2.

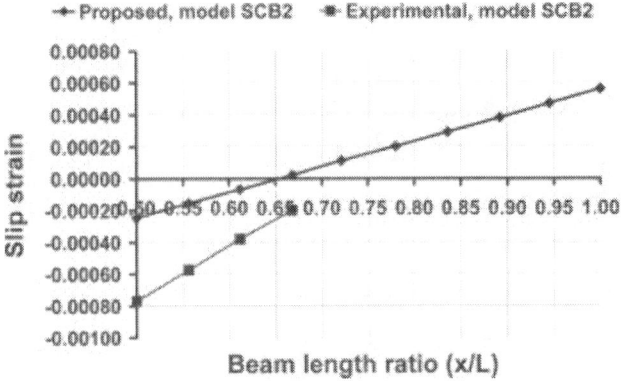

Figure 17b. Interface slip strain values of the lower slab for beam model SCB2.

Guide to Stability Design Criteria for Metal Structures | 237

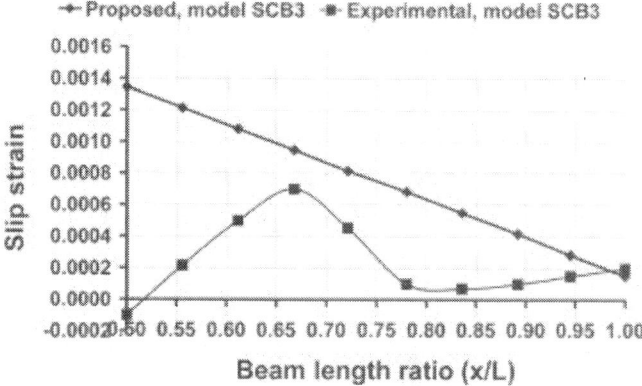

Figure 18a. Interface slip strain values of the upper slab for beam model SCB3.

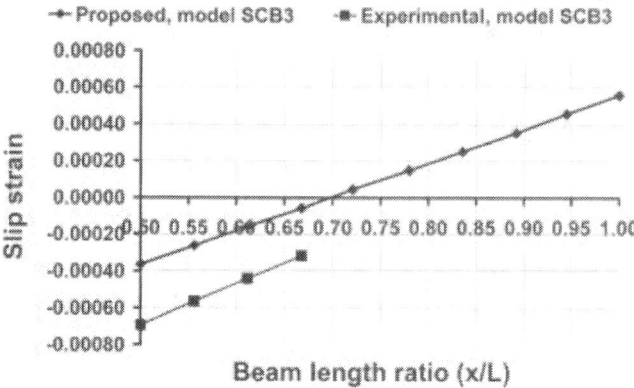

Figure 18b. Interface slip strain values of the lower slab for beam model SCB3.

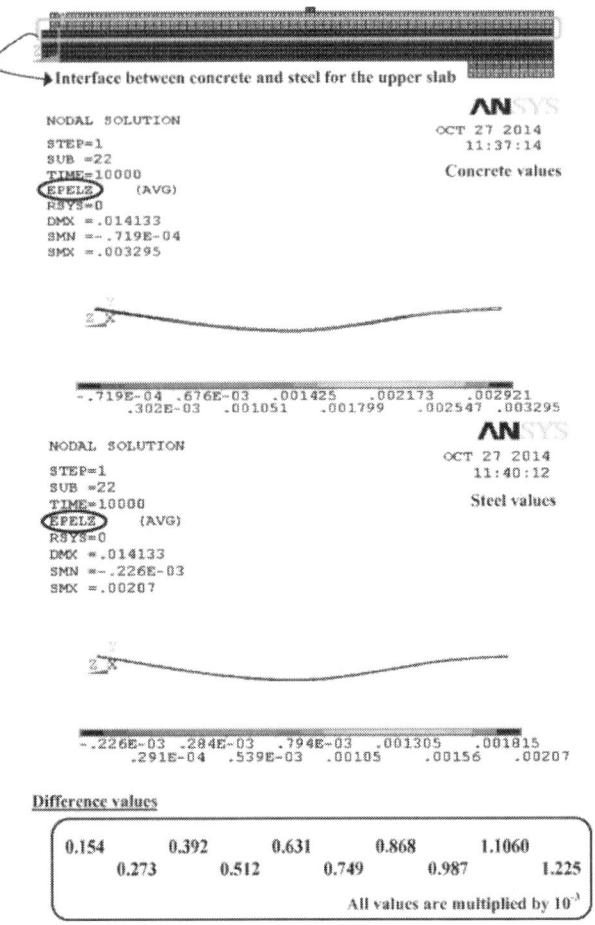

Figure 19. Difference between the interface longitudinal strains of concrete and steel along the beam length direction for the upper slab of the beam model SCB1 (slip strain values).

PARAMETRIC STUDIES

To further improve the understanding of the strength capacity and the fracture behavior of the continuous double steel-concrete composite beams having head studs shear connectors, parametric studies were performed to investigate the impact of the presence or absence of lower slab at the interior support, and the variation of the steel beam height. In addition, the variation of the lower slab length and thickness, and the variation of the studs arrangement and diameter are also studied.

The Influence of Removing the Lower Slab

The case study under consideration involves the influence of removing the lower slab on the mechanical and geometrical characteristics of the beam models at failure, such as the strength and the deflection capacity values. The study was conducted on three proposed models SB1, SCB2, and SCB3 respectively. Fig. 20 illustrates the effect of varying composite action on the fracture characteristics (strength and maximum deflection) of the proposed model.

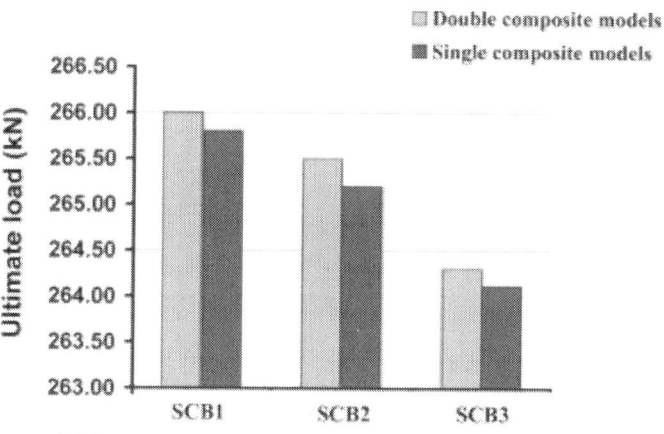

a. Ultimate load as a function of the composite action

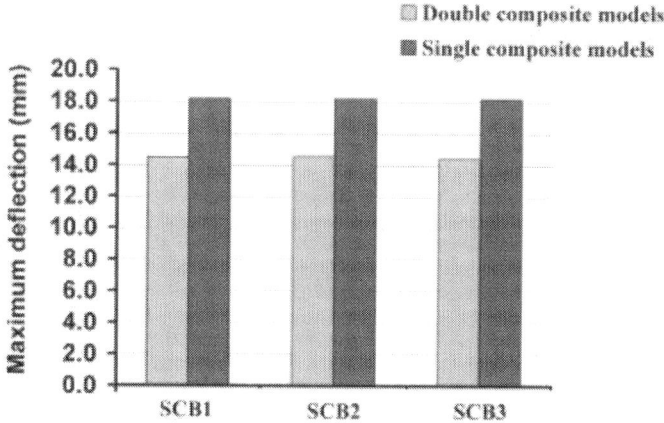

b. Maximum deflection as a function of the composite action

Figure 20. Fracture characteristics as a function varying composite action.

Fig. 20.a compares the results obtained for the ultimate load values, taking into account the presence or absence of the lower slab. It has to be noted that in case of existing the lower slab, the proposed ultimate load values increase by an amount of 0.075% in the case of model SCB1, and by an amount of 0.11% in the case of model SCB2. This increasing value is become 0.068% in the case model SCB3. One can observe that the presence of the lower slab increases the strength capacity by an average amount 0.08% for all experienced composite models. This means that the existence of the lower slab has a minor effect on the strength capacity values.

Fig. 20.b presents a comparison of the results of the maximum deflection values. Again, the above three models were investigated twice in order to experience the effect of removing the lower slab. It has to be noted that in case of existing the lower slab, the proposed values of the maximum deflection decrease by a significant average amount of 20% for all beam models.

The Influence of Varying the Steel Beam Height

In this part, the effect of changing the height of steel beam on the characteristics of the collapse stage for the continuous double steel-concrete composite beam is investigated. Five steel beam heights of values 110 mm, 130 mm, 150 mm, 170 mm, and 190 mm were proposed and applied to the model SCB1, as a case study. Fig. 21 demonstrates the effect of varying steel beam height on the fracture characteristics of the proposed model.

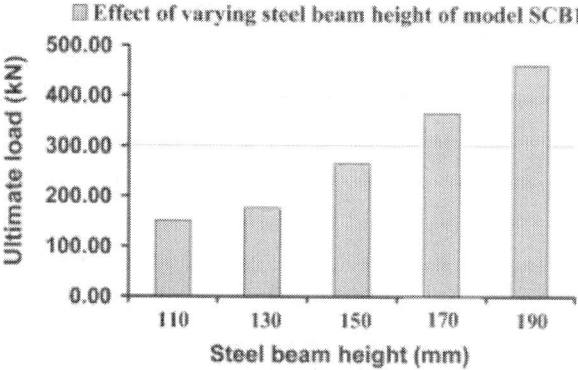

(a) Ultimate load as a function of the steel beam height of model SCB1

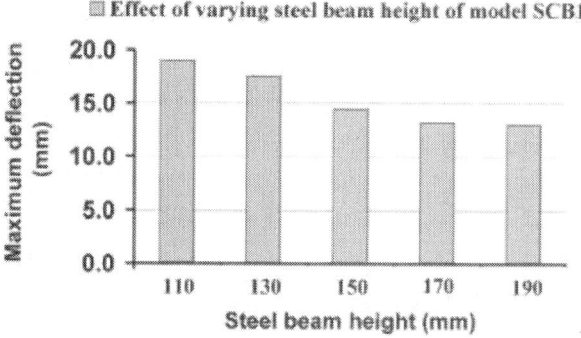

(b) Maximum deflection as a function of the steel beam height of model SCB1

Figure 21. Fracture characteristics as a function varying steel beam height of model SCB1.

Fig. 21.a presents a comparison of the results of the ultimate load values. This study was applied to the model SCB1 with the same height values as indicated previously. It has to be noted that the case of the beam model with steel beam height of 110 mm had the minimum ultimate load value, whereas the case of the steel beam height of 190 mm had the maximum one. The increase in the ultimate load for two consecutive heights (e.g. 130 mm and 150 mm) reached a significant value of approximately 30% for all models.

Fig. 21.b compares the results obtained for the maximum deflection values, taking into account the same model and the proposed steel beam heights as indicated above. It has to be observed that the case of the 110 mm steel beam height had the maximum value of the maximum deflection, whereas the case of the steel beam with height of 190 mm had the minimum one. The decrease in the maximum deflection values for two consecutive heights (e.g. 130 mm and 150 mm) reached a significant value of approximately 12% for all models.

The Influence of Varying Lower Slab Length

This part contains study of the impact of changing the lower slab length on the collapse stage characteristics for the continuous double steel-concrete composite beam. Four lower slab lengths of values 1000 mm, 1200 mm, 1400 mm, and 1600 mm were proposed and applied to the model SCB1, as a case study. Fig. 22 exhibits the effect of varying the lower slab length on the fracture characteristics of the proposed model.

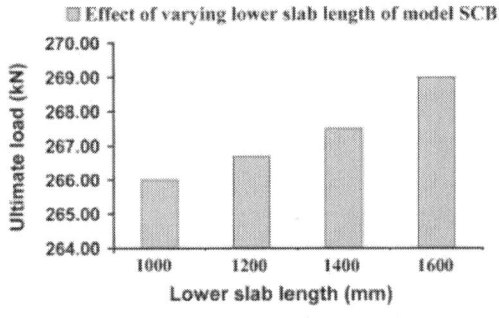

(a) Ultimate load as a function of the lower slab length of model SCB1

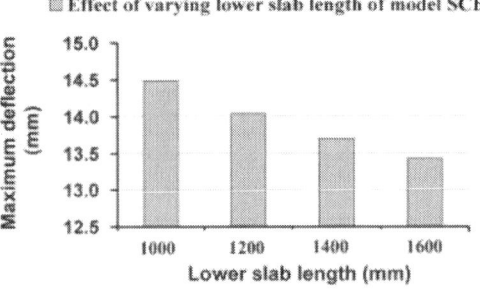

(b) Maximum deflection as a function of the lower slab length of model SCB1

Figure 22. Fracture characteristics as a function varying lower slab length of model SCB1.

Fig. 22.a presents a comparison of the results of the ultimate load values. This study was applied to the model SCB1 with the same lower slab length values as mentioned above. It has to be observed that the case of the beam model with lower slab length of 1600 mm had the maximum ultimate load value, whereas the case of the lower slab length of 1000 mm had the minimum one. The increase in the ultimate load for two consecutive slab lengths (e.g. 1200 mm and 1400 mm) reached non-notable value of approximately 0.6% for all models.

Fig. 22.b compares the results obtained for the maximum deflection values, taking into account the same model and the proposed lower slab lengths as indicated above. It has to be noted that the case of the beam model involving lower slab length of 1600 mm had the minimum value of the maximum deflection, whereas the case of the lower slab length of 1000 mm had the maximum one. The decrease in the maximum deflection values for two consecutive slab lengths (e.g. 1200 mm and 1400 mm) reached a remarkable value of approximately 5% for all models.

The Influence of Varying Lower Slab Thickness

The influence of changing the lower slab thickness on the characteristics of the collapse stage for the continuous double steel-concrete composite beam is studied herein. Four lower slab thicknesses of values 80 mm, 100 mm, 120 mm, and 140 mm were proposed and executed to the model SCB1, as a case study. Fig. 23 explicates the effect of varying the lower slab thickness on the fracture characteristics of the proposed model.

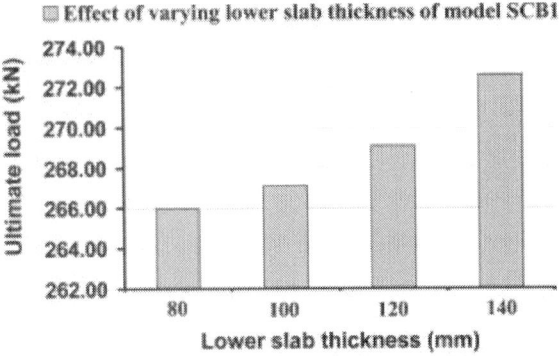

(a) Ultimate load as a function of the lower slab thickness of model SCB1

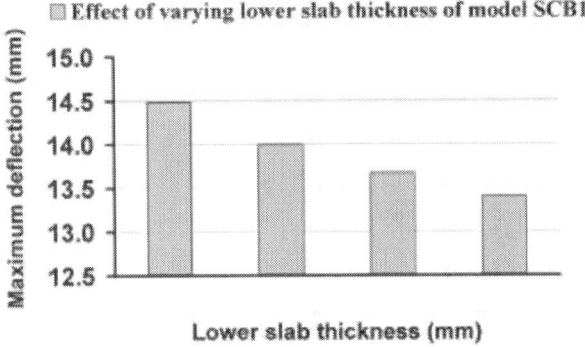

(b) Maximum deflections as a function of the lower slab thickness of model SCB1

Figure 23. Fracture characteristics as a function varying lower slab thickness of model SCB1.

Fig. 23.a presents a comparison of the results of the ultimate load values. This study was applied to the model SCB1 with the same lower slab thickness values as mentioned above. It has to be noted that the case of the beam model with lower slab thickness of 80 mm had the minimum ultimate load value, whereas the case of the lower slab thickness of 140 mm had the maximum one. The increase in the ultimate load for two consecutive slab thicknesses (e.g. 100 mm and 120 mm) reached non-remarkable value of approximately 0.85% for all models.

Guide to Stability Design Criteria for Metal Structures | 245

Fig. 23.b compares the results obtained for the maximum deflection values, taking into account the same model and the proposed lower slab thicknesses as mentioned above. It has to be observed that the case of the beam model including lower slab thickness of 80 mm had the maximum value of the maximum deflection, whereas the case of the lower slab thickness of 140 mm had the minimum one. The decrease in the maximum deflection values for two consecutive slab thicknesses (e.g. 100 mm and 120 mm) reached a slightly remarkable average value of approximately 3.5% for all models.

The Influence of Varying the Head Studs Arrangement

In order to complete the parametric study, the effect of changing the arrangement of the head studs on the characteristics of the collapse stage for the continuous double steel-concrete composite beam is discussed. Three cases of head studs arrangement were proposed. The first case is when the studs were fully arranged along the whole length of the upper and the lower interface slab-steel beam surfaces. The second case is when the studs were arranged with staggered shape, while the third case is when the studs were completely removed. This study was applied to the model SCB1, as a case study. Fig. 24indicates the effect of varying the studs arrangement on the fracture characteristics of the proposed model.

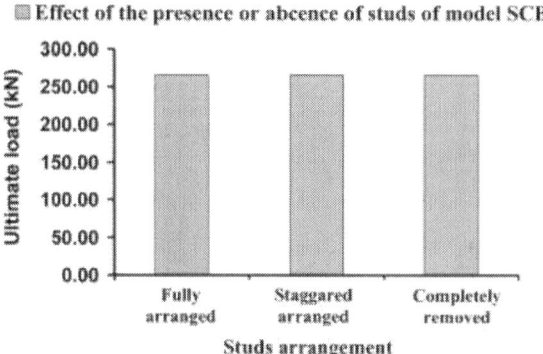

(a) Ultimate load as a function of the studs arrangement of model SCB1

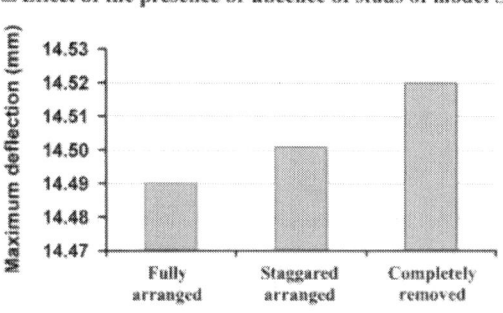

(b) Maximum deflection as a function of the studs arrangement of model SCB1

Figure 24. Fracture characteristics as a function varying studs arrangement of model SCB1.

Fig. 24.a presents a comparison of the results of the ultimate load values. This study was applied to the model SCB1 with the same cases of the studs arrangement as mentioned above. It has to be noted that the change of the shape of the studs arrangement has no influence on values of the ultimate load.

Fig. 24.b compares the results obtained for the maximum deflection values. It has to be observed that the beam model including fully studs arrangement had the minimum value of the maximum deflection, whereas the case of the completely removed head studs had the maximum one. The increase in the maximum deflection values for two consecutive studs arrangement (e.g. fully and staggered arrangement) reached a very slightly remarkable value of approximately 0.08% for all models.

The Influence of Varying the Head Studs Diameter

The effect of changing the value of the diameter of the head studs on the characteristics of the collapse stage for the continuous double steel-concrete composite beam is examined as a part of this study. Four cases of head studs diameter of values 13 mm, 16 mm, 19 mm, and 22 mm were suggested and implemented to the model SCB1, as a case study. Fig. 25 clarifies the effect of varying the studs diameter on the fracture characteristics of the proposed model.

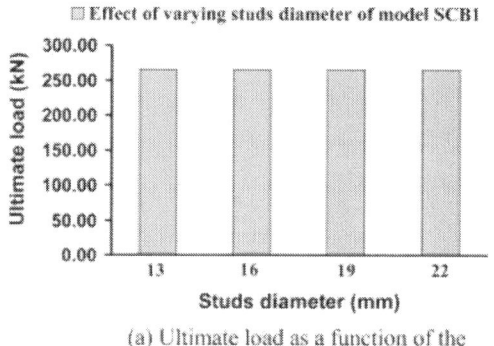

(a) Ultimate load as a function of the studs diameter of model SCB1

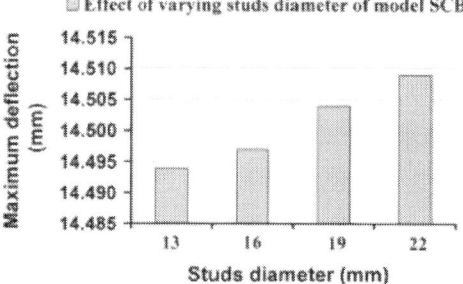

(b) Maximum deflection as a function of the studs diameter of model SCB1

Figure 25. Fracture characteristics as a function varying studs diameter of model SCB1.

Fig. 25.a presents a comparison of the results of the ultimate load values. This study was applied to the model SCB1 with the same values of the head studs diameters as stated above. It has to be noted that the change of the head studs diameter has no influence on values of the ultimate load.

Fig. 25.b compares the results obtained for the maximum deflection values. It has to be concluded that the beam model containing 13 mm head studs diameter had the minimum value of the maximum deflection, whereas the case of the studs diameter of 22 mm had the maximum one. The increase in the maximum deflection values for two consecutive studs diameter (e.g. 16 mm and 19 mm) attained a very slightly notable average value of approximately 0.05% for all models.

CONCLUSIONS

This paper investigates the behavior of the continuous steel-concrete composite beam taking into account the existence of the double composite action and the head stud shear connectors.

Based on the finite element numerical study and the experimentally available results, the following main conclusions can be extrapolated:

1. A numerical proposed model based on the finite element theory can be used to examine the geometrical and mechanical characteristics in steel-concrete composite beam with double composite action, resulting in a good agreement when comparing to available full-scale test data.
2. The comparison of the strength capacity values obtained by the proposed and experimental models leads to a good agreeable between them. An average increase in the proposed strength capacity values of approximately 29% compared to the experimentally available data was concluded for all proposed models. However, a softer performance of the validation figures (load – deflection curves) is observed for the developed models than that of the experimental results. This is mainly due to the varying range of the William-Warnke constants values, which must be chosen carefully

by a sensitivity analysis in order to encounter the Ansys convergence problems as mentioned above.
3. An increase in the proposed interface steel-concrete slip values of approximately 38% compared to the experimentally available data was observed, leading to slightly non-agreeable results.
4. An increase in the proposed interface steel-concrete slip strain values of approximately 49% and 55% compared to the experimentally available data was observed for both the upper and the lower slabs respectively, leading to somewhat non-agreeing values between them. This is due to the difference in values of friction (shear slip) at the slab-steel beam interface between the analytical and experimental approaches, because of the presence of the contact elements for simulating this friction. This means that the shear slip has a significant contribution to composite beam deformation, which cannot be negligible.
5. Parametric studies were carried out to look at the impact of removing the lower slab, the effect of varying steel beam height and the lower slab length and thickness, and the effect of changing the head studs arrangement and diameter. These studies were performed to investigate the effect of these parameters on the strength and the deflection capacity of the steel-concrete composite beams having double composite action.
6. The presence of the lower slab increases the proposed strength capacity values by an average amount 0.08% for all experienced composite models, leading to a minor effect on the strength capacity. Moreover, the proposed values of the maximum deflection decrease by a significant average amount of 20% for all beam models when removing the lower slab.
7. In comparison with the five suggested cases of steel beam height involved in the parametric study, it can be observed that the more increase the steel beam height is the bigger the ultimate load values are.
8. Moreover, this study showed that the smaller the lower slab length or thickness is the smaller the ultimate load values and the bigger the maximum deflection values are.
9. It can be noted that the change of the shape of the studs arrangement has no influence on the values of the ultimate load. In addition, the beam model including fully studs arrangement had a

minimum value of the maximum deflection, whereas the case of the completely removed head studs had the maximum one.
10. In comparison with the five head stud diameters suggested in this study, one can concluded that the change of this parameter has no effect on the values of the ultimate load. In addition, it has to be noted that the smaller the head studs diameter is the smaller the maximum deflection values are.

REFERENCES

1. Rozsas A. Plastic design of steel–concrete composite girder bridges. M.sc thesis, department of structural engineering, faculty of civil engineering: Budapest (Hungary); 2011.
2. Xu C, Su Q, Wu C. Experimental study on double composite action in the negative flexural region of two-span continuous composite box girder. J Constr Steel Res 2011;67(10):1636–48.
3. Tan El, Uy B, Hummam G. Behavior of multi-span composite steel–concrete beams subjected to combined flexure and torsion. Research and applications in structural engineering, mechanics and computation, London, UK; 2013. p. 1397–402.
4. Lin W, Yoda T. Numerical study on horizontally curved steel– concrete composite beams subjected to hogging moment. Int J Steel Struct, USA 2014;14(3):557–69.
5. Henriques D, Goncalves R, Gamotim D. Nonlinear analysis of steel–concrete beams using generalized beam theory. In: 11th World Congress on Computational Mechanic, Barcelona, Spain; 20–25 July 2014.
6. Liang Q, Uy B, Bradford M, Ronapf H. Ultimate strength of continuous composite beams in combined bending and shear. J Constr Steel Res 2004;60(8):1109–28.
7. Liang Q, Uy B, Bradford M, Ronapf H. Strength analysis of steel–concrete composite beams in combined bending and shear. J Struct Eng, ASCE, USA 2005;131(10):1593–600.
8. Sebastian W, McConnel RE. Nonlinear finite element analysis of steel–concrete composite structures. J Struct Eng, ASCE, USA 2000;126(6):662–74.
9. Hirst MJS, Yeo MF. The analysis of composite beams using standard finite element programs. Comput Struct 1980;11(3):233–7.
10. Al-Amery RIM, Roberts TM. Nonlinear finite difference analysis of composite beams with partial interaction. Comput Struct 1990;35(1):81–7.
11. Salari MR, Spacone E, Shing B, Frangopol DM. Nonlinear analysis of composite beams with deformable shear connectors. J Struct Eng, USA 1998;124(10):1148–58.

12. Thevendran V, Chen S, Shanmungam NE, Liew JWR. Nonlinear analysis of steel–concrete composite beams curved in plan. Finite Elem Anal Des 1999;32(3):125–39.
13. Reiner S. Bridges with double composite action. Struct Eng Int, UK 1999;1:32–6.
14. Stroh SL, Sen R.** Steel bridge with double composite action: innovative design. In: 5th International bridge engineering conference, tampa, FL (US). Transportation Research Record, April 3–5, 2000. vol. 1, 1696. p. 299–309.
15. Newmark NM, Siess CP, Viest IM. Test and analysis of composite beams with incomplete interaction. Proc, Soc Exp Stress Anal 1951;9:75–92.
16. Duan S, Niu R, Xu J, Zheng H. A finite element model for double composite beam. Challenges, opportunities and solutions in structural engineering and construction, London; 2010. p. 197– 202.
17. Duan SJ, Huo JH, Zhou QD. The research on calculation method of the ultimate bearing capacity of double steel– concrete composite beam. J Shijiazhuang Railway Inst 2007;20(4):1–4.
18. Duan SJ, Zhou QD, et al. Experimental study on bearing capacity of double steel and concrete composite continuous beams. J Railway Sci Eng 2008;5(5):12–7.
19. Nagai M, Inaba N, et al. Experimental study on ultimate strength of composite and double composite girders. In: Proceedings of 8th Pacific structural steel conference steel structures in natural hazards, 2007. p. 329–34.
20. Duan SJ, Duan YJ, Zhang ZG. The interface slip expression of double steel–concrete composite beam under concentrated load. J Shijiazhuang Railway Inst 2007;20(2):1–4.
21. Yang XW, Duan SJ. The effective width of reinforcement bars for double steel–concrete composite beam. Eng Mech 2008;25(A1): 184–8.
22. Wang G, Wang FJ, et al. Theoretical analysis of double composite beam deformation in elastic state by Goodman elastic sandwich method. Chin Railway Sci 2006;27(5):66–70.
23. Yen BT, Huang T, et al. Steel box girders with composite bottom flanges. In: Official proceedings, 3rd annual international bridge conference. Pittsburgh, PA (US); 1986. p. 79–86.
24. Duan SJ, Wang JW, Zhou QD, Wang HL. An experimental study on double steel–concrete composite beam specimens. Challenges, opportunities and solutions in structural engineering and construction, London; 2010. p. 209–14.
25. Fanning P. Nonlinear models of reinforced and post tensioned concrete beams. Lecture, Department of Civil Engineering University College Dublin Earls fort Terrace. Dublin, Ireland; 2001.

26. William KJ, Warnke ED. Constitutive model for the triaxial behavior of concrete. Proc of the Int Assoc Bridge Structural Engineering, ISMES, Bergamo 1975;19:174.
27. Razaghi J, Hosseini A, Hatami F. Finite element method application in nonlinear analysis of reinforced concrete structures. Second Nat Congr Civil Eng 2005.
28. Kachlakev D, Miller T. Finite element modeling of reinforced concrete structures strengthened with FRP laminates. Oregon state University; May 2001.
29. Wolanski J, B.S. Flexural behavior of reinforced and prestressed concrete beams using FRP element Analysis, A thesis submitted to the faculty of the graduated school, Marquetee university, in partial fulfillment of the requirement for the degree of master of science; May 2004.

CITATION

Ashraf Mohamed Mahmoud, Finite element modeling of steel concrete beam considering double composite action, Ain Shams Engineering Journal, Available online 4 May 2015, ISSN 2090-4479, http://dx.doi.org/10.1016/j.asej.2015.03.012.

Chapter 7

Protection of Steel Corrosion in Concrete Members by the Combination of Galvanic Anode and Nitrite Penetration

Minobu Aoyama[1], Shinichi Miyazato[2], and Mitsunori Kawamura[3]

[1]Department of Maintenance Engineering, Central Nippon Highway Engineering Nagoya Co Ltd, Kanazawa Branch, EkinishiHonmacthi, Kanazawa, Ishikawa 920-0025, Japan
[2]Department of Civil and Environmental Engineering, Kanazawa Institute of Technology, Hakusan, Ishikawa 924-0838, Japan
[3]Kanazawa University, Kakumamachi, Kanazawa, Ishikawa 920-1192, Japan

INTRODUCTION

The effectiveness of nitrite compounds as an admixture toprevent steel bars from corroding in concrete containingchloride ions is generally well documented [1–5]. However,there is still little attempt to prevent the corrosion of steel barsby the penetration of nitrite ions fromthe outside to hardenedconcrete with relatively large amounts of chloride ions exceptthat Ngala et al. [3] assessed the efficiency of calcium nitriteas an inhibitor when used in surface treatment.

We found many RC slabs with cracks which had beencaused by externally supplied chloride-induced corrosion ofsteel bars. The slabs were partly exfoliated on their surfaces.In these concrete members, steel bars were arranged perpendicularto each other in near-surface regions. Steel barssituated on the surface side severely corroded, but those onthe inner side were not so significantly damaged. In the repairworks of such damagedRCmembers, spaces left behindwhenconcrete with large amounts of chloride up to the depthof corroded steel bars was removed have been filled withsufficient amounts of nitrite-containing mortars. The repairmethod is based on the expectation that supply of nitrite ionsfrom the mortars can inhibit corrosion of steel bars situatedat inner positions. However, the movement of nitrite ionsthrough hardened concrete is very slow. Hence, it takes longtime for sufficient amounts of nitrite ions to move from themortars to old concrete. Then, we proposed the combinedcorrosion protection method of the nitrite penetration andthe galvanic anode. Namely, in the method, a galvanic anodeprocess was applied to steel bars by the use of zinc wires untilsufficient amounts of nitrite ion reached the steel bars situatedat inner positions.This composite method has an advantagethat we can economically apply it to reinforced concretemembers with large amounts of chloride ions brought fromthe outside over long times. However, it should be noted thatgood workability was required for the mortars when mixedwith large amounts of a corrosion inhibitor. From this pointof view, we used lithium nitrite as a corrosion inhibitor. Inthis paper, various experimental results to give assurance forthe realization of this method were provided. Moreover, theactual execution of works and the specific design method tobe implemented for the application of this technical methodare also described.

CHARACTERISTICS OF CORROSION PROGRESS OF STEEL BARS

Corrosion Condition of Steel Bars in the Bridge Slabs

Thecorrosion of steel bars in the bridge slabs in the expresswaywas caused by the penetration of chloride deicers. Figure 1shows changes in chloride ion concentration with the depthfrom surfaces in the vicinity of the under face of deteriorateddecks at the edge of the RC bridge. Concrete covers wereexfoliated in places around the portions. As seen in thisfigure, chloride ion concentration had reached the highestvalue around steel bars in most bridge slabs. Frominfiltrationconditions of chloride ions in the bridge slabs, steel barsin concrete were under severe saline atmospheres even inthe inner portions. However, corrosion concentrated on steelbars in the surface side; those in the inner portions onlyslightly corroded, as shown in Figure 2.

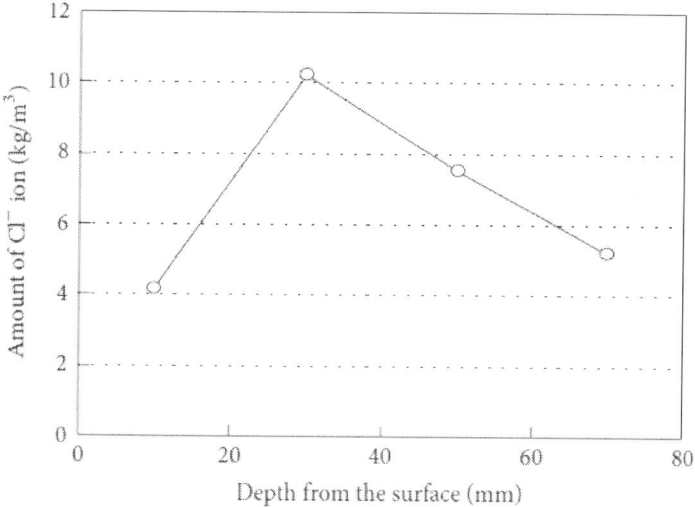

Figure 1: Distribution of chloride ion concentration in the regions near the under surface of RC bridge slabs deteriorated by the penetrationof deicing salt.

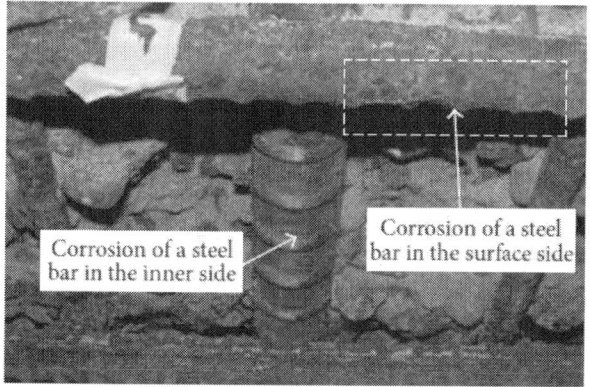

Figure 2: Actual situations of corrosion of steel bars in the vicinity of the under surface in a bridge slab.

Experiments for Confirming the Formation of Macrocells in Steel Bars

In order to clarify mechanisms of differences in the degree of corrosion between steel bars in the surface and the inner side in near-surface regions in the bridge slabs, an experiment was conducted simulating the actual situations of steel bars in the process of degradation [6] (Figure 3). After mortar specimens (water/cement rate = 0.5) were submitted to immersion in NaCl solution-drying repetitions, the total corrosion current density passing through the steel bars in the specimens was measured.

As shown in the upper part in Figure 3, steel bars (13mm in diameter) were halved in the longitudinal direction and then embedded in mortar specimens (125 × 125 × 35 mm) so as to perpendicularly intersect one another. Concrete cover in no. 1, no. 2, and no. 3 specimens is 5mm thick and in no.4 it is 0 mm. In order to produce various stages of the degree of corrosion of steel bars, in no. 3 specimen, a slit of 0.2mm in width is made in concrete cover, and, in no. 4 specimen, steel bars were directly exposed to corrosive conditions. Faces in concrete prisms except the under face (125 × 125 mm)(Figure 3) were sealed with the epoxy resin, and then the concrete prisms underwent immersion in NaCl solution drying repetitions for 3 months (immersion in 3% NaCl solution for 3 days, exposure to an 60%R.H. atmosphere for 4days at 20°C). At a week after the completion of immersion in NaCl solution-drying repetitions, the total corrosion current density between each element in steel bars was measured.

Figure 4 shows the maximum values of the total corrosion current density between each part in divided steel bars at various corrosion states. From this figure, progression rate of corrosion at the upper half in steel bars is different from that at the lower half. As seen in Figure 4, in the specimen no.4 without concrete cover and the specimen no. 3 with a slit in concrete cover, corrosion rapidly proceeded in steel bars situated on the surface side but slowly did in the other part($H\,②, N\,①, N\,②$). However, large amounts of chloride ions were present in concrete around these steel bars. These results suggest that macrocells composed of the half of a steel bar in the surface side as the anode ($H\,①$) and the other half and the inner steel bars as the cathode ($H\,②, N\,①, N\,②$) were formed. In bridge slabs damaged by externally supplied chloride, half of steel bars situated on the surface side are supposed tobecome the anode; the other half of the steel bars and steelbars in the inner side are supposed to become the cathode.As a result, the half of steel bars as the galvanic anode situatedon the surface side severely corroded. Namely, large amountsof chloride ions have penetrated into deep portions in bridgeslabs, but the degree of corrosion in steel bars situated in theinner portions was slight.

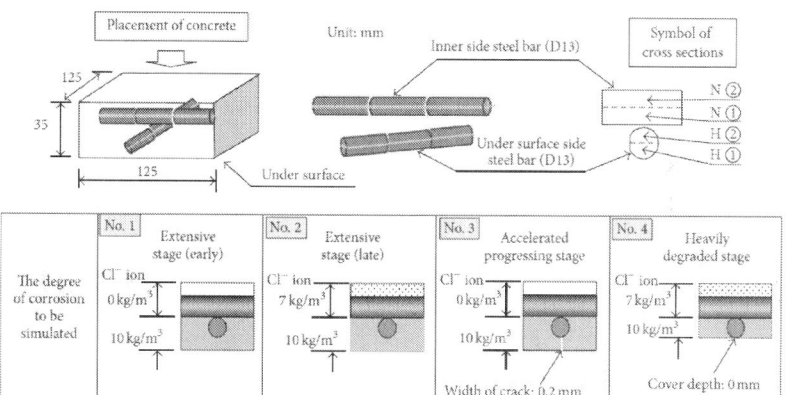

Figure 3: Measurements of current density in steel bars with progressing degree of corrosion.

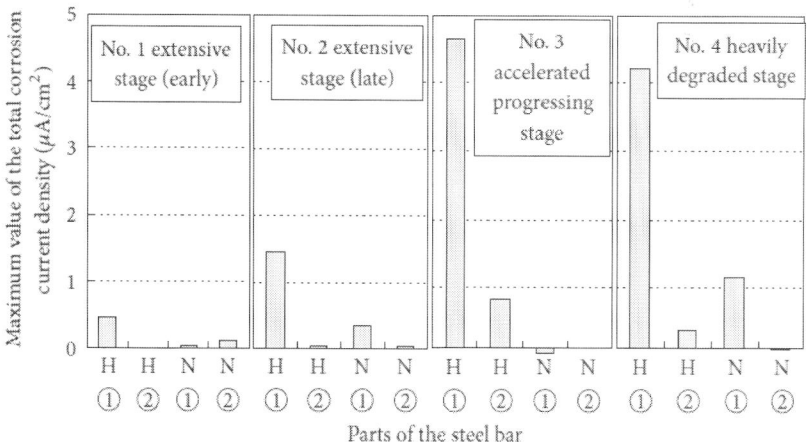

Figure 4: Average value of the total corrosion current density between each divided part in steel bars at various stages of corrosion state.

EFFECTIVENESS OF CORROSION INHIBITOR-CONTAINING MORTAR LAYERS

Effects of Lithium Nitrite on the Corrosion of Steel Bars and the Movement of NO_2^- Ions from Mortars to Concrete.

In order to apply the combined corrosion control methodto actual concrete structures damaged by steel corrosiondue to chloride ions externally supplied, it is important toconfirm whether externally supplied nitrite ions have corrosioninhibitive effects on steel bars in chloride-containingconcrete. It is also necessary to quantitatively examine acritical NO_2^-/Cl^- molar ratio to suppress the corrosion ofsteel bars in the concrete.

In this study, the accelerated indoor and outdoor exposuretests were conducted to discuss these subjects [7]. Asshown in Figure 5, in the former exposure tests, 200 × 400× 150mm concrete specimens were used; in the latter 200 ×800 × 150mm ones, NaCl was added to concrete to simulateexternally supplied chloride-containing concrete. Concretespecimens (W/C = 63%, C = 210 kg/m^3) were composed oftwo parts with low and high Cl^- ion contents of 2.5 kg/m^{3}and 7 kg/m^3 (Figure 5). In both of the exposure tests, twotypes of specimens with and without a corrosion inhibitor(LiNO$_2$) were used, and in the other specimen, a thin mortarlayer 20mm thick containing a large amount of LiNO$_2$ of57.5 kg/m^3 was applied to the half part of top faces in acorrosion inhibitor-free concrete specimen. The molar ratioOf NO_2^-/Cl^- in the LiNO$_2$ and NaCl-containing concrete was 0.8. Steel bars were embedded in near-surface regions on the top faces.

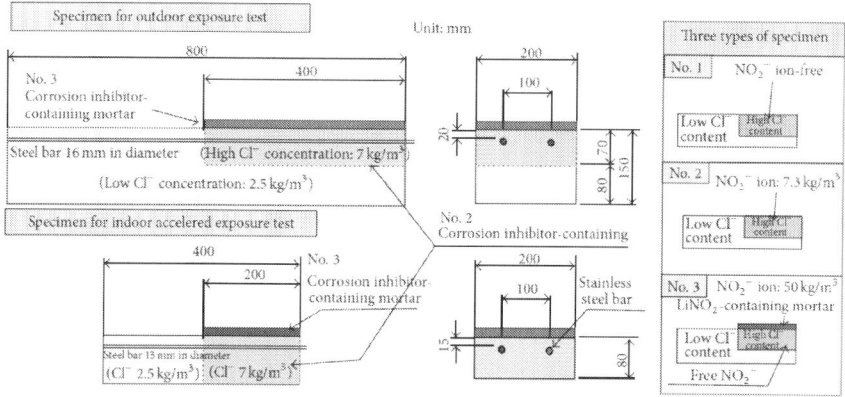

Figure 5: Specimens for the outdoor and the indoor exposure tests.

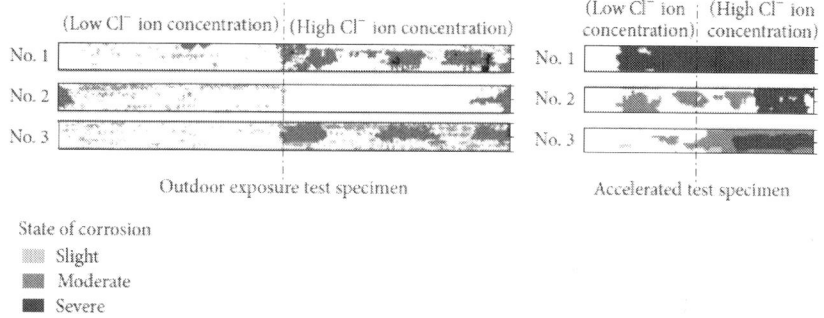

Figure 6: States of corrosion on steel bars after the completion of exposure tests.

The large specimens were exposed to outdoor environments for 37 months; the small ones wetting-drying repetitions (drying in 50% R.H. for 4 days and wetting in >95% R.H. for 3 days at 40°C) for 10 months in the laboratory.

Figure 6 shows the states of corrosion of steel bars in concrete after the exposure tests. It is found from this figure that, in LiNO$_2$-containing concrete specimens with a large amount of Cl$^-$ ion (NO$_2^-$/Cl$^-$ molar ratio of 0.8) (no. 2 specimen), corrosion of steel bars is considerably suppressed as compared to inhibitor-free specimens. Moreover, a comparison in the state of corrosion between no. 1 and no. 3 specimen in the accelerated tests clearly indicates that the addition of LiNO$_2$ suppressed the corrosion of steel bars even in concrete with a large amount of Cl$^-$ ion. In particular, itshould be noted that the application of a thin mortar layerwith a large amount of $LiNO_2$ to the top faces of concretespecimens (no. 3 specimen) suppressed the corrosion of steelbars. This fact suggests that NO_2^- ions penetrated from themortar into concrete to considerable extent. In fact, judgingfrom a distribution of NO_2^- ion concentration in the depthdirection measured in no. 3 specimen exposed to outdoorenvironments (Figure 7), we can prove that NO_2^- ions in themortar gradually penetrated into the concrete.

From the test results stated above, it is certain that thereplacement of concrete with mortars containing NO_2^- ionscould suppress the corrosion of steel bars surrounded byold concrete with large amounts of Cl^- ions. Furthermore,corrosion inhibitive atmospheres will be formed around steelbars in concrete members in corrosive saline environmentswhen a thin mortar layer with a large amount of $LiNO_2$ isapplied to their surfaces.

Quantitative Evaluation of Permeability of NO_2^- Ion.

In order to design the repair works by the applicationof a thin mortar layer with a large amount of corrosioninhibitor to surfaces of hardened concrete including Cl^- ions,it is necessary to quantitatively evaluate the permeability ofNO_2^- ion from themortars to concrete.

Outdoor exposure tests were conducted using 200 ×130 × 500mm concrete prisms for quantitatively evaluatingthe penetration rate of NO_2^- ion from $LiNO_2$-containingmortar layers to hardened concrete. As shown in Figure 8,concrete prisms with $LiNO_2$-containingmortar layers 20mmthick on the top faces were exposed to natural environmentsabout 5km from the coast for 37 months. The mixproportions of concrete and mortar used in the exposuretests are water/cement ratio of 50% (unit cement content =342 kg/m³) and 45% (sand/cement ratio = 2.5), respectively.Steel bars 16mm in diameter were embedded to simulateactual reinforced concrete. Two types of mortars containingchloride and nitrite ions of 50 kg/m³ were used in mortarlayers on the top faces of concrete prisms for comparing thepenetration rate of NO_2^- and Cl^- ions. Actually, as shown inFigure 8, mortars layers 20mm thick were applied on a halfof the top face of a concrete specimen (200mm × 500 mm).After that, other five faces of concrete specimens except underfaces were sealed with the epoxy resin. Only the under facesof concrete prisms were exposed to natural environments for37 months. Figure 9 shows NO_2^- and Cl^- ion profiles at theend of exposure tests. As seen in this figure, both the ionsin mortar layers gradually intruded into concrete. It is alsofound from this figure that NO_2^- ions in the mortars morerapidly moved into concrete than Cl^- ions did.

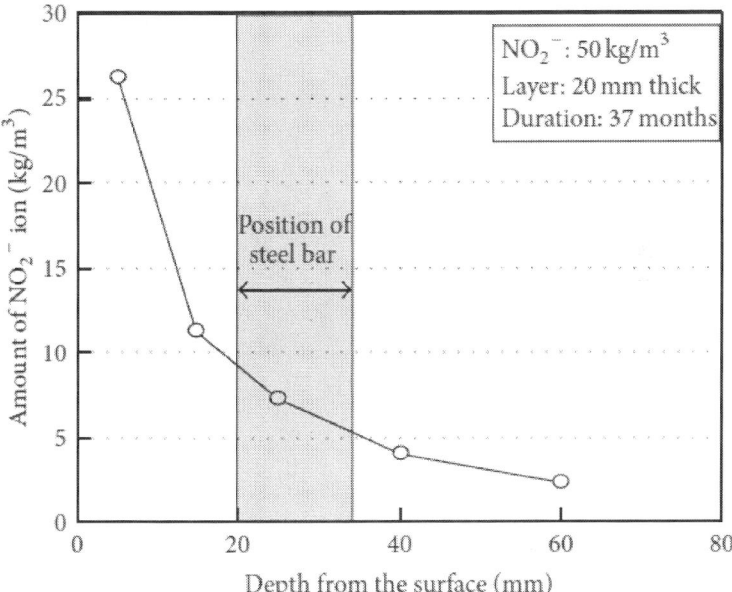

Figure 7: Distribution of NO_2^- ion concentration in the concretespecimen attached with a thin mortar layer containing a largeamount of $LiNO_2$ exposed to outdoor environments.

Values of apparent diffusion coefficient (D) were calculatedby the finite difference method using NO_2^- and Cl^- ioncontents in concrete measured at various depths at the end ofexposure tests (37 months). Five faces except the under facewere sealed by the epoxy resin. Hence, the total amounts ofNO_2^- and Cl^- ions within concrete specimens did not changeduring and even after the exposure tests. D values of NO_2^- and Cl^- ion calculated using the profiles of both the ions inthe concrete prisms (Figure 9) were $4 \times 10^{-8} cm^2/s$ and $3.3 \times 10^{-8} cm^2/s$, respectively.Thus, D value of NO_2^- ion is higherthan that of Cl^- ion.Therefore, even if the diffusion coefficientofNO_2^- ion is assumed tobe the same as thatofCl^- ion in theprediction of penetration of NO_2^- ions from mortars layersinto concrete in the design of practices, the assumption is onthe safe side.

CONFIRMATION OF THE EFFECTIVENESS OF ZINC WIRE AS A GALVANIC ANODE

Exposure Tests

As shown in Figure 10, supposing therepair of concretemembers in a saline environment, the specimensin which zinc wires as a galvanic anode were arrangedalong steel bars were exposed to accelerated exposure tests.The corrosion mitigation effect of the zinc wires was evaluatedby measuring the total corrosion current density passingthrough steel bars [8]. Steel bars in concrete specimens (200× 150 × 350 mm) were halved in the longitudinal direction.Two steel bars (13mm in diameter) situated on the exposedface side and the other one (19mm in diameter) in the innerportions were embedded to perpendicularly intersect eachother. The thickness of concrete cover of the specimens is0mm.Thechloride ion content in concrete up to 13mmfromthe exposed face was 15 kg/m³; that in the inner portions137mm thick 10 kg/m3. Five faces except the under face (theexposed face of 200 × 350 mm) in specimens were sealedwith the epoxy resin as shown in Figure 10. These threespecimens were exposed to a wetting-drying repetitions (1week) (immersion in water for 3 days, drying in 50% R.H.atmosphere for 3 days and wetting in 90% R.H. for 1 days at40°C). After the completion of 8 wetting-drying repetitions,the concrete specimens were placed in an atmosphere at 40°Cfor 1 week, and then the total corrosion current densitiesbetween each element of steel bars were measured.

One specimen was the control. The other two concreteprisms were used for seeing the effects of $LiNO_2$-containingmortar layers on the corrosion protection of steel bars andthe effectiveness of zinc wires as a galvanic anode. Namely,concrete portions 13mm thick in which steel bars wereembedded, were removed, and then a zinc (ϕ 2 mm) wire as agalvanic anode was arranged along steel bars in the specimenno. 3.Thereafter, the spaces occupied by old concrete (25mmthick) were filled with a mortar containing a large amountof LiNO$_2$ (55 kg/m^3). These three specimens were exposedto the same accelerated wetting-drying repetitions tests for8 weeks as the untreated original concrete specimens.

At one week after the completion of the exposure tests,the total corrosion current densities between each element ofsteel bars were measured.

Figure 11 shows themaximumvalue of the total corrosioncurrent densities between each element in steel bars in specimensbefore and after the application of LiNO$_2$-containingmortar layer and a zinc galvanic anode.Themaximum valuesof the total corrosion current densities in this figure referto the maximum value among the ones obtained in eachelement of steel bars in divided parts. In Figure 11, it isseen that very large values of the total corrosion currentdensity were measured for all of the three specimens withouttreatments after the completion of accelerated exposure tests(8 weeks). Namely, severe corrosion was in progress in steelbars embedded in the specimens. As seen in Figure 11, atthe completion of the first accelerated exposure tests for8 weeks, corrosion is found to actively progress on the upper half (N ②) in the inner steel bars. In particular, inno. 1 specimen without treatments (the control) (Figure 10),very large corrosion current densities are measured on thelower half of the steel bars on the exposed face side (H①) after the completion of both the first and the secondaccelerated exposure testing (16 weeks). Figure 11 indicatesthat the corrosion current densities both on the upper andlower half of steel bars in the exposed face (H ①, H ②) in no. 2 and no. 3 specimens) are greatly reduced by thetreatments with LiNO$_2$-containing mortars and zinc wires.This result indicates that the removal of Cl$^-$ ions containingportions at the end of the first accelerated exposure test for 8weeks established corrosion protection environments aroundthe steel bars. Furthermore, another important indicationobtained from Figure 11 is that the corrosion current densityin the lower half of steel bars in the inner side (N ①) wasreduced after the application of mortar layers with LiNO$_2$.Thus, it is certain that NO$_2^-$ ions moved from the mortarlayers into the old concrete.

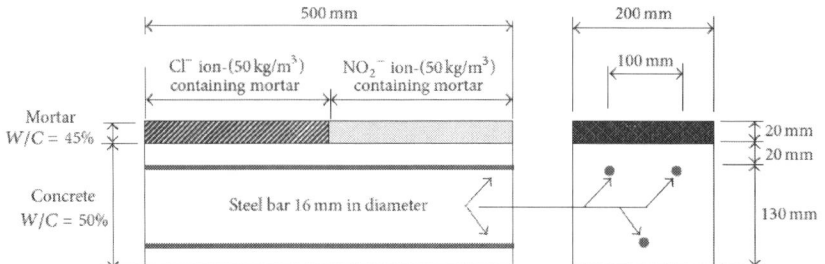

Figure 8: Concrete prisms for determining the penetration rate of NO_2^- and Cl^- ions from mortars to concrete.

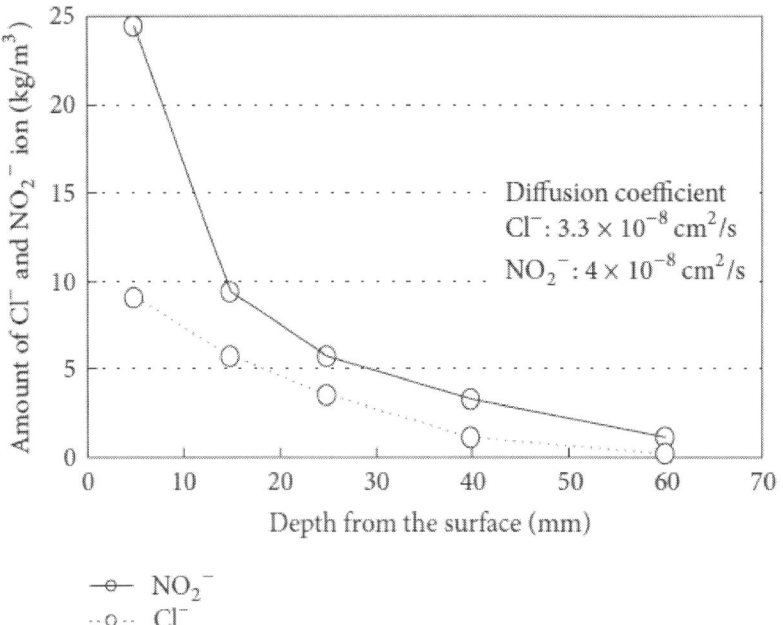

Figure 9: NO_2^- and Cl^- ions profiles in concrete at the end of exposure tests.

The corrosion current density on the upper half of steel bars in the inner portions in no. 2 specimen (N②) decreased compared with that in the specimen without treatments. But the value of the corrosion current density itself is great. It is presumed that sufficient corrosion protection environments have not been established around steel bars. On the other hand, in specimen no. 3 with an attached zinc wire, the corrosion current densities in steel bars in both the upper and the lower half of the steel bars (N ①, N ②) in the inner portions are smaller than those in specimen no. 2. In particular, the value of corrosion current density in the upper half in the steel bars in the inner portions in specimen no. 3 (N ②) is a third that in specimen no. 2. These results show that effects of a zinc wire as a galvanic anode appear at early stages.

In conclusion, it was confirmed that the arrangement of zinc wires along the steel bars as a galvanic a node was effective for protecting steel bars from corrosion in concrete structures in corrosive salt environments.

4.2. Conditions for Zinc Wires to Function as a Galvanic Anode. The formation of passivation films on the surface of zinc wires should be avoided for them to function as a galvanic anode. It is an essential condition that pHof the pore solution in concrete around steel bars is maintained above 13.3 [9]. Lithium nitrite is used as a corrosion inhibitor in the present corrosion protection method. The pH value in pore solutions in $LiNO_2$-containing concrete increases with increasing amounts of the corrosion inhibitor. Pore solutions were extracted from mortar cylinder (ϕ 50 × 100 mm) containing an amount of $LiNO_2$ of 55 kg/m³ cured for 24 hours. Figure 12 shows a relationship between NO_2^- ionconcentration and pH value in the pore solutions. Assuming that pH value is proportional to NO_2^- ion concentration in the pore solution in small ranges, the addition of about 40 kg/m³ asNO_2^- ions to themortars is required for pHvalue to be held about 13.3.

It is assured that the process of the formation of corrosion products around zinc wires in concrete is not expansive.

PRACTICES IN THE COMPOSITE CORROSION PROTECTION METHOD

Overview of the Method

By using zinc wires as a galvanic anode, we can protect steel bars from corrosion even in the concrete structures exposed to corrosive Saline environments since the completion of repair works. Thus, the extent of concrete to be removed can be limited to ranges of depths of cracks.

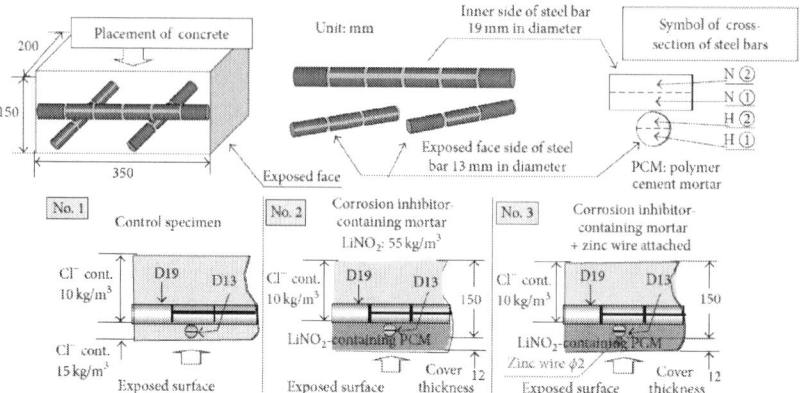

Figure 10: Specimens of exposure tests for confirming the effectiveness of zinc wire as a galvanic anode.

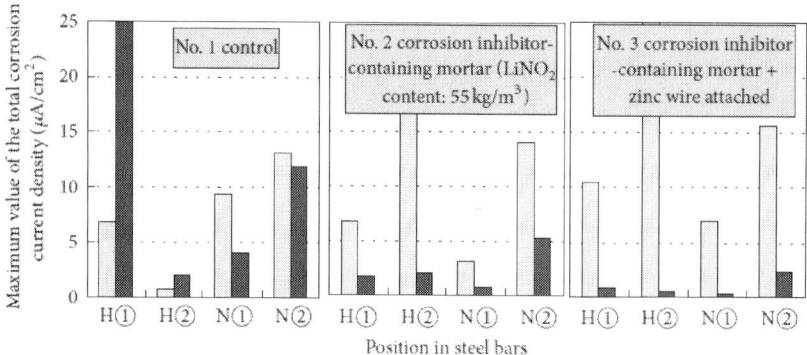

Figure 11: Maximum values of the total corrosion current densities between elements of steel bars before and after the application of $LiNO_2$-containing mortar layers and a galvanic anode.

However, zinc wires used as a galvanic anode dissolve with time. Hence, the time for the zinc wires to effectively function as a galvanic anode is limited. On the other hand, in the corrosion protection method by the application of corrosion inhibitor-containing mortar layers, it takes long times for $NO2^-/Cl^-$ molar ratio in atmospheres around steel bars to be higher than 0.8. Therefore, the combined method of galvanic anode and nitrite penetration is based on the concept that the advantages of the exhibition of the function as a galvanic anode immediately after the application and the assurance of passivating atmospheres around steel bars for long periods by the penetration of nitrite are combined. A flow chart of the design in the present method is given in Figure 13. It is possible to minimize the total amount of repair and renewal cost. Thus, this method can be applied to concrete members reaching accelerated corrosion or heavily damaged stages in which concrete covers have partly exfoliated. Thus, even if we apply this method to heavily damaged concrete members, steel bars in concrete will be passivated for long periods. Figure 14 shows a photograph of a model made for better understanding of details of the combined corrosion protection technique. Firstly, zinc wires (2~4mm in diameter) were arranged along steel bars (10mm in diameter) and then connected to the inner steel bars on the upper side at a space of 125mm to 300mm by spot welding. After that, steel bars (19mm in diameter) intersecting at right angles are covered by a large amount of LiNO$_2$-containing mortar layers with 15~20mm thick. Spaces above the LiNO$_2$-containing mortar layers are filled with a corrosion inhibitor-free mortar for usual repair works. In relation to the use of a large amount of LiNO$_2$-containing mortar, it was confirmed by experiments thatmortarswith a LiNO$_2$ content of 70 kg/m^3 were workable enough to do practical works for filling portions removed in damaged concrete members.

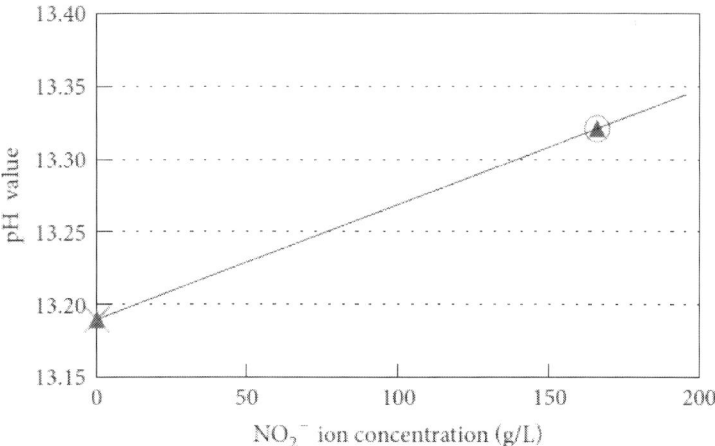

Figure 12: Relationship betweenNO$_2^-$ ion concentration and pH inpore solutions in mortars.

Details of Design

As shown in Figure 13, firstly, the depth to be chipped, the amount of corrosion inhibitor, and the thickness of corrosion inhibitor-containingmortar layers are laid down. Secondly, the time up to the formation of an atmosphere for assuring the passivation of steel bars in existing concrete members is estimated. Finally, we judge the applicability of the present method to a damaged concrete member by verifying whether a zinc wire will function as a galvanic anode up to the attainment of the passivation of steel bars by nitrite ions.

The definite procedures are as follows.

1) Values of C_0 (chloride ion concentration on the surface) and D (apparent diffusion coefficient) are estimated from chloride ion concentration distributions in concrete.
2) By using the C_0 and the D values, the movement of Cl^- ions, which remain behind in old concreteportions, and NO_2^- ions in a large amount of $LiNO_2$- containingmortar layers can be predicted by the finite difference method.
3) It is possible to estimatewhen steel bars are passivated based on theNO_2^-/Cl^- molar ratio.
4) Effective period of zinc wires as a galvanic anode can be estimated fromreductions in themass of zinc wires by corrosion.
5) Reductions in the mass of zinc wires are estimated from the value of corrosion current density in the steel bars after repair treatments using Faraday's law of electrolysis (see (1)):

$$W = i \cdot t \cdot \frac{m}{2F}, \qquad (1)$$

where W is the corrosion weight loss per unit surface area (g/cm^2), i is the corrosion current density (A/cm^2), t is the corrosion time (s),$m$is the molecular weight of zinc (65.38 g/mol), and F is the faradayconstant (96500 C/mol).

6) The life time (L) of zinc wires is calculated by the use of (2). It is considered that the steel bars havebeen passivated at the end of the effective period.It is assumed that both the annual corrosion weightloss and the corrosion current density are zero atthat time. Both the corrosion current density andthe annual corrosion weight

loss are also assumed toproportionally decrease time. A safety factor shouldbe taken into consideration in actual calculations:

$$L = \frac{Z}{(Fs \times As \times \Delta W/2)}, \qquad (2)$$

where Z is the mass of zinc (g), Fs is the safety factor, As is the total surface area of steel bars (cm²), and$\Delta W$is the annual corrosion weight loss of zinc wires afterapplication (g/cm²/year).

If it is confirmed from the calculations that sufficient amounts of NO_2^- ion can penetrate into old concrete during the effective period of zinc wires as a galvanic anode, the combined method of corrosion protection will be applicable.

A Case Study of the Application

The case study is concerned with a degraded part of a 35-year-old RC bridge slab damaged by spraying of deicing salts (Figure 15).Thenominal design strength and an apparent diffusion coefficient (D) of the original concrete are 24N/mm² and 3.0×10^{-8} cm²/s, respectively.

The details of steel bars used in the parts to be repaired are as follows.

Steel bars on the surface side in the under face of the slab are 25mmin diameter and spaced 125 mm; those on the inner side are 32mm in diameter and spaced from 100 to 200 mm; shear steel bars are 16mm in diameter and spaced 125mm. Concrete cover is 35mm thick. Concentration distributions of chloride ions in near-surface regions on the under face side are provided in Figure 16. It is found fromthis figure that the chloride ion concentration at a position of steel bars is about 7 kg/m³.

Details of Repair
A schematic diagram of the parts to be repaired is given in Figure 17. Deteriorated concrete parts 55mmthick were removed. As a result, steel bars on the inner side in the under face were disclosed.We attached zinc wires (3mm in diameter) on steel bars (10mm in diameter), as shown in Figure 17. The portions removed were restored by applying a LiNO₂-containing mortar layer (LiNO₂ content: 70 kg/m³), 20mm thick, and a usual mortar layer, 35mm thick.

Prediction of the Time up to the Attainment of a Passivating Condition

Distributions of Cl^- and NO_2^- ionconcentration immediately after repair and those predicted by the finite difference method are provided in Figure 16. A diffusion coefficient of 0.2×10^{-8} cm^2/s was used for both types of mortars in the analysis.

Figure 18 shows changes in Cl^- and NO_2^- ion content, NO_2^-/Cl^- molar ratio with time at a depth of 85mm (the position of inner steel bars). It is found from this figure that the time up to the attainment of a critical NO_2^-/Cl^- molarratio of 0.8 for passivating steel bars was 6.7 years.

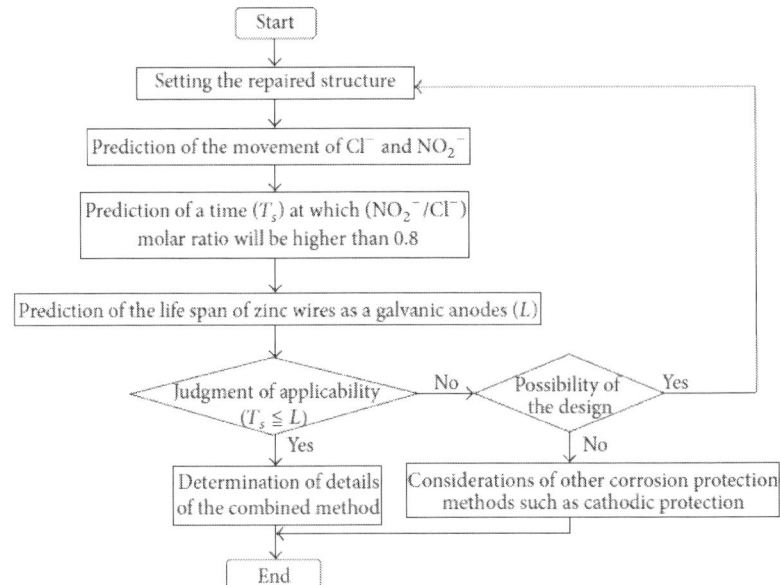

Figure 13: Flow chart of the design.

Guide to Stability Design Criteria for Metal Structures | 271

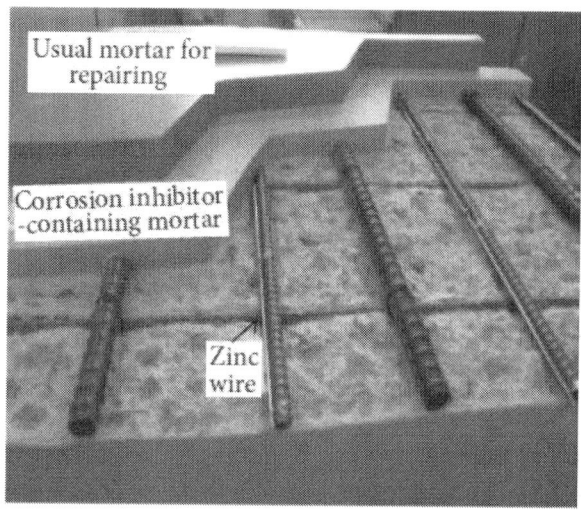

Figure 14: Model of details of the combined corrosion protectiontechnique.

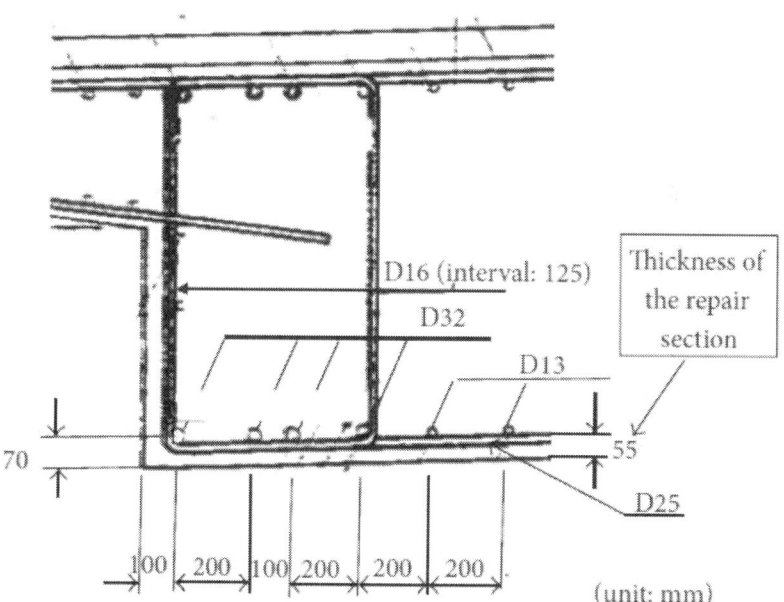

Figure 15: Part of the RC bridge slab.

PRACTICES IN THE COMPOSITE CORROSION PROTECTION METHOD

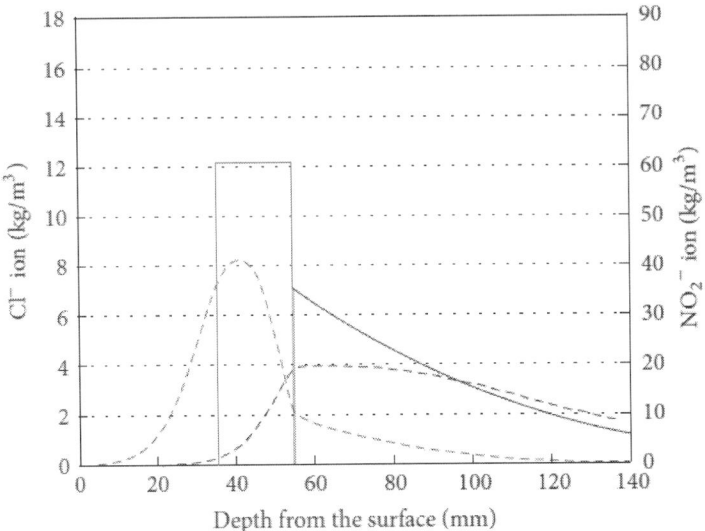

— Cl⁻ ion immediately after repairing
--- Cl⁻ ion at 7 years
— NO₂⁻ ion immediately after repairing

Figure 16: Cl⁻ and NO₂⁻ ion concentration profiles immediatelyafter repairing and their profiles predicted up to 7 years.

Prediction of the Life Time of Zinc Wires

A zinc wirewith 1m long has a charge of surface area of approximately0.17m2 of steel bars to be passivated. Assuming that thecorrosion current density of steel bars immediately afterrepair is 0.3 $\mu A/cm^2$ (corrosion current density of a low tomoderate level), a calculationby the use of (1) gives an annualcorrosion weight loss of zinc of 0.0033 g/cm²·year. The lifetime of zinc wires L (years) in the case of arranging zinc wireswith a diameter of 3mm(0.505 g/cm) can be calculated by theuse of (2). Assuming that the safety factor is 2,

$$L = \frac{0.505}{2} \times \left(1700 \times \left(\frac{(0.0033 + 0)}{2}\right)\right) = 9 \text{ years}. \quad (3)$$

As described above, the time up to the formation of anatmosphere for passivating steel bars was 6.7 years. Hence,the combined corrosion protection method is applicable tothe repair of this deteriorated bridge slab.

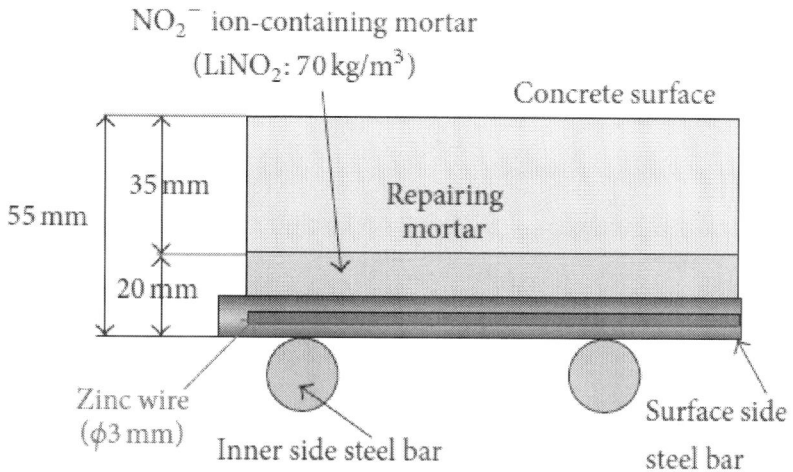

Figure 17: Schematic diagram of the parts repaired.

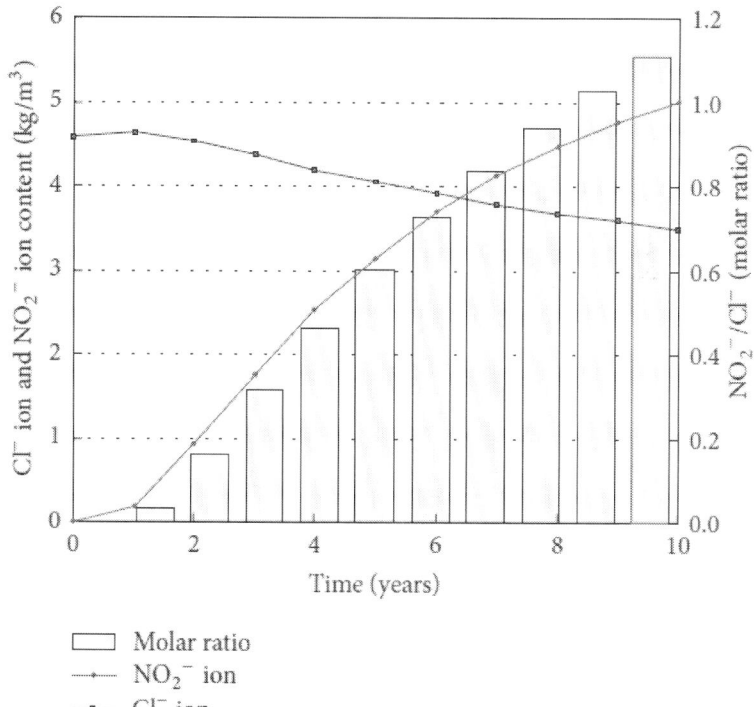

☐ Molar ratio
─■─ NO_2^- ion
─●─ Cl^- ion

Figure 18: Changes in Cl^- and NO_2^- ion content, NO_2^-/Cl^- molarratio with time at a depth of 85mm (the position of inner steel bars).

PRACTICES IN THE COMPOSITE CORROSION PROTECTION METHOD

Practical Works of Repair

A flow chart of the practicalworks of repair is provided in Figure 19. Cracked andexfoliated portions were removed with the hand breaker.However, in order to prevent concrete around steel bars frombeing damaged and to remove rust on the steel bars, it ispreferable to conduct the chipping work by the water jetmethod. Chipping had been conducted until surfaces of steelbars on the inner side were disclosed.

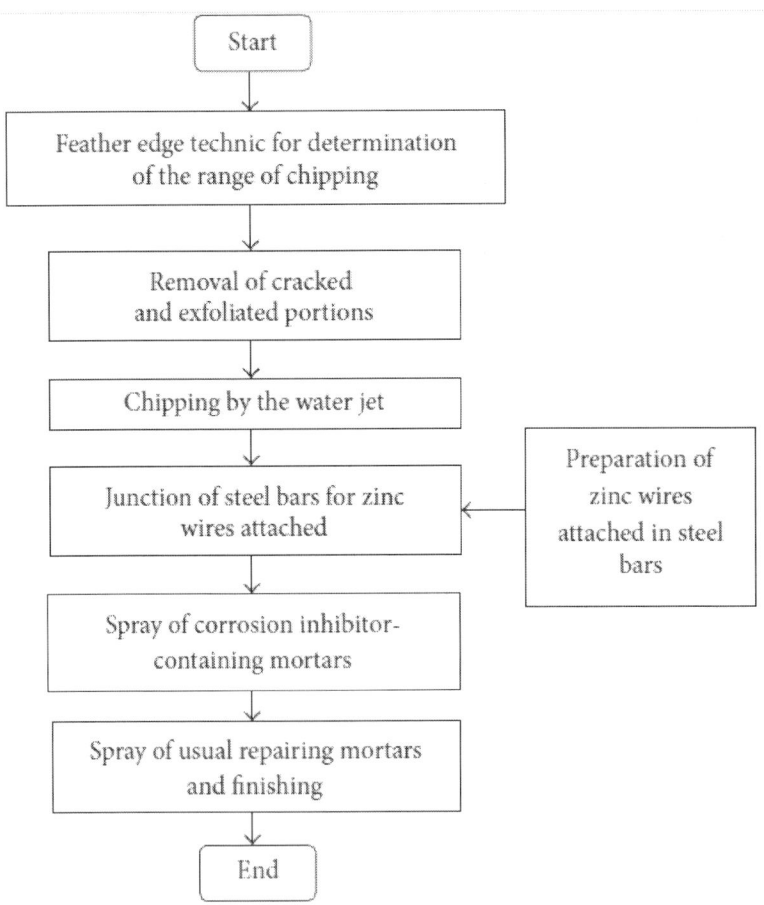

Figure 19: Flow chart of the practical works of repair.

Guide to Stability Design Criteria for Metal Structures | 275

Figure 20: State of a zinc wire bound to steel bars.

In advance, zinc wires were bound to steel bars 10mm indiameter at intervals of 200mmin a factory and then installedin the construction site. Surfaces of steel bars attached to zincwires should be ground and polished in advance. We placeda zinc wire along steel bars on the straight and tightly boundboth using a mild steel wire (the electric resistance betweensurfaces: 0Ω) with diameter of about 0.8mm (Figure 20). It is important tomake sure that the electric resistance betweenthe bound mild steel wires, steel bars and the zinc wires isless than 0.3Ω.

In order to secure good galvanic electricity between zincwire and steel bars, both were bound by spot welding atintersecting spots. To fix zinc wires around the central sectionof a corrosion inhibitor containing-mortar layer with a highNO$_2^-$ ion concentration, zinc wires are positioned from5mm to 10mm above the top of steel bars on the inner side,and then on the opposite side a zinc wire was welded to steelbars (see Figure 21).

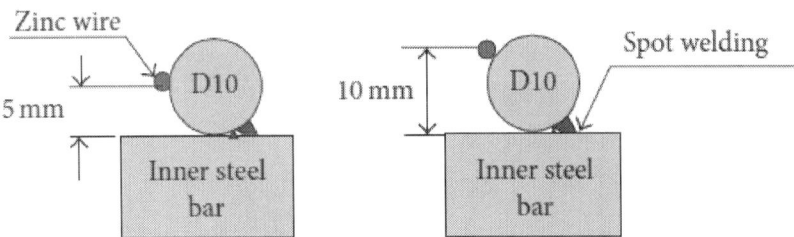

Figure 21: Position of a zincwire attached.

SUMMARY AND CONCLUSIONS

The applicability of a corrosion protection method by thecombination of galvanic anode and nitrite penetration toreinforced concrete members damaged by the chlorideinducedcorrosion of steel bars were discussed. The majorresults obtained are summarized as follows.

1. Many concrete structures have been damaged by externally supplied chloride-induced corrosion of steel bars. In these cases, it was proven by experiments that the corrosion was caused by the macro cells formed between the half of steel bars on the surface side (anode) and the other half and steel bars in the inner portions (cathode).
2. A large amount of $LiNO_2$ containing mortar layers applied to the surfaces of concrete members can supply concrete around steel bars with sufficient amounts of NO_2^- ions to prevent corrosion during the effective period of zinc wires as a galvanic anode.
3. In a case study on a bridge slab damaged by externally supplied chloride-induced corrosion of reinforcement, the applicability of the combined method to the slabs was proven by calculating the amount of NO_2^- ions penetrated during the life time of zinc wires.

Conflict of Interests

The authors declare that there is no conflict of interestsregarding the publication of this paper.

REFERENCES

1. 1 J. Tritthart and P. F. G. Banfill, "Nitrite binding in cement,"Cement and Concrete Research, vol. 31,no. 7, pp. 1093–1100, 2001.
2. 2 M. Kawamura, S. Tanikawa, R. N. Swamy, and H. Koto, "Poresolution composition and electrochemical behavior of steel barsin mortars with nitrite corrosion inhibitors," in Proceedings ofthe 5th CANMET/ACI International Conference on Superplasticizersand Other Chemical Admixtures in Concrete, pp. 35–53,Rome, Italy, 1997.
3. 3 V. T. Ngala, C. L. Page, and M. M. Page, "Corrosion inhibitorsystems for remedial treatment of reinforced concrete.Part 1:calciumnitrite," Corrosion Science, vol. 44, no. 9, pp. 2073–2087,2002.
4. 4 K. Y. Ann, H. S. Jung, H. S. Kim, S. S. Kim, and H. Y.Moon, "Effect of calcium nitrite-based corrosion inhibitor inpreventing corrosion of embedded steel in concrete," Cementand Concrete Research, vol. 36, no. 3, pp. 530–535, 2006.
5. 5 N. S. Berke and M. C. Hicks, "Predicting long-term durabilityof steel reinforced concrete with calcium nitrite corrosioninhibitor," Cement and Concrete Composites, vol. 26, no. 3, pp.191–198, 2004.
6. 6 H. Hirano, S. Miyazato, K. Yamamoto, and M. Takenouchi,"The explanation of salt damage mechanism progress on RCroad bridge by spraying antifreeze agent," Proceedings of JapanConcrete Institute, vol. 29, no. 1, pp. 1005–1010, 2007 (Japanese).
7. 7 M. Aoyama, S. Hirano, D. Asae, and K. Torii, "Corrosionprotection of steel bars in concrete in salt damage due to corrosioninhibitor mixing mortar," Proceedings of Japan ConcreteInstitute, vol. 27, no. 1, pp. 931–936, 2005 (Japanese).
8. 8 T. Ishikawa, S. Miyazato, M. Aoyama, and S. Hirano, "Developmentof repair method that combines the galvanic anodematerial and corrosion inhibitor for salt damage of the RC,"Repair of Concrete Structures, vol. 10,pp. 29–36, 2010 (Japanese).
9. 9 A. Bentur, S. Diamond, and N. S. Berke, Steel Corrosion inConcrete, E & FN Spon, London, UK, 1992.

CITATION

Minobu Aoyama, Shinichi Miyazato, and Mitsunori Kawamura, "Protection of Steel Corrosion in Concrete Members by the Combination of Galvanic Anode and Nitrite Penetration," International Journal of Corrosion, vol. 2014, Article ID 618280, 11 pages, 2014. doi:10.1155/2014/618280

Index

A
Accelerated electrochemical technique (ACET), 188
Aggregate skeleton, 1, 3, 4, 17
Aggressive aqueous environment, 158
Allowable stress design (ASD), 73
Anodic polarization current, 146
Apparent diffusion coefficient, 261, 268, 269

B
beam height, 213, 238, 240, 241, 242, 249

C
Cathodic polarization, 101, 103, 104, 107, 108, 134, 135, 138, 143, 144, 145, 146, 147
Chloride-induced corrosion, 103, 106, 110, 111, 116, 125, 126, 127, 131, 132, 147
Chromate conversion coating (CrCC), 177
Coefficient of variation (COV), 77
Combined corrosion protection method, 254, 272
composite plate girder bridge, 215
composite structure, 214, 250
Compressive Packing Model (CPM, 27
Compressive stresses, 11, 12, 13, 15
concrete construction, 213
concrete section, 214
Controlled low-strength material (CLSM), 2
Controlled low-strength materials (CLSM), 1, 23
Corrosion criterion, 111
corrosion inhibitor, 254, 258, 260, 265, 277
Corrosion inhibitor, 102, 103, 148, 266, 268, 275
Corrosion resistance, 102
Critical pigment volume concentration (CPVC), 184
cross-sectional area, 214
Current density, 103, 143, 144, 145, 146

D

Diameter (Dm), 41, 42, 51
Distilled water, 104, 105, 115, 134

E

Electrochemical impedance spectroscopy (EIS), 183, 202
Electrochemical Impedance Spectroscopy (EIS), 186

F

Flade potential (EF), 108
Friction coefficient, 70, 72, 78, 80, 81, 82, 85, 86, 89, 92, 93, 98

I

Inductive coupled plasma (ICP), 182
Inflatable structures, 69, 72, 73, 74, 82, 93, 96, 97, 98, 100
Intermetallic (IM), 152

L

Limit states design (LSD), 73
Local electrochemical impedance spectroscopy (LEIS), 194
lower slab thickness, 213, 220, 243, 244, 245

M

Mixes manufactured, 21
mortar layer, 258, 260, 261, 263, 269
Mortar layer, 275

N

Near surface deformed layer (NSDL), 152

Nitrite compounds, 253
Nitrite concentration, 101, 102, 103, 112, 113, 115, 120, 122, 123, 124, 125, 126, 131, 132, 143, 144, 145, 146, 147
Noticeable influence, 15

P

Passivation ability, 108
Passive oxide, 104, 149
pH value, 107
Pigment volume concentration (PVC), 184
plastic design, 215
Polarisation resistance, 163, 186
Potentiodynamic polarisation (PP), 183, 186
pure steel, 213

R

RC bridge, 255, 269, 271

S

saline environments, 260
Saline environments, 265
Schematic diagram, 269
Selective ion electrode technique (SIET), 194
Self-compacting concrete (SCC), 25, 26
Staircase potentio-electrochemical impedance spectroscopy (SPEIS), 190
Steel concrete composite, 213
steel section, 214, 219
steel-concrete composite beam, 213, 216, 217, 218, 220, 228, 238, 249

Stern-Geary constant, 101, 103, 104, 135, 143, 144, 145, 146, 147
Structural engineering material, 151
Sulfate threshold level, 126, 131, 132, 147

W

Water film thickness (WFT), 29